APPLICATIONS
of HIGH
TEMPERATURE
POLYMERS

APPLICATIONS
of HIGH
TEMPERATURE
POLYMERS

edited by
Robert R. Luise

CRC Press
Taylor & Francis Group
Boca Raton London New York

CRC Press is an imprint of the
Taylor & Francis Group, an **informa** business

First published 1997 by CRC Press
Taylor & Francis Group
6000 Broken Sound Parkway NW, Suite 300
Boca Raton, FL 33487-2742

Reissued 2018 by CRC Press

Library of Congress Cataloging-in-Publication Data

Applications of high temperature polymers / edited by Robert R. Luise.
    p.   cm.
    Includes bibliographical references and index.
    ISBN 0-8493-7672-6
    1. Polymers. 2. Heat resistant materials. I. Luise, Robert R.
TA455.P58A68 1996
620.1'9204217—dc20   96-8586

A Library of Congress record exists under LC control number: 96008586

Publisher's Note
The publisher has gone to great lengths to ensure the quality of this reprint but points out that some imperfections in the original copies may be apparent.

Disclaimer
The publisher has made every effort to trace copyright holders and welcomes correspondence from those they have been unable to contact.

ISBN 13: 978-1-315-89067-8 (hbk)
ISBN 13: 978-1-351-06977-9 (ebk)

Visit the Taylor & Francis Web site at http://www.taylorandfrancis.com and the
CRC Press Web site at http://www.crcpress.com

PREFACE

This book is based mainly on the proceedings of a conference organized and chaired by the editor on High Temperature Polymers: Advances and Applications held February 20 and 21, 1995 in Clearwater Beach, Florida, and sponsored by Executive Conference Management, Inc.

The conference represented a gathering by leading experts in the rapidly growing field of high temperature polymers to discuss recent advances, applications, and marketing projections. Speakers/authors represent a diverse group composed of academicians, industrial researchers, consultants, managers, and marketing forecasters. Papers cover timely topics related to liquid crystalline polymers; polyimides; and other important high temperature organic polymers, including chemistry and key functional properties in moldings, films, fibers, and coatings as well as applications in electronics, packaging, and friction/wear.

The editor wishes to thank the authors for their contributions, Executive Conference Management for its cooperation, and the editors at CRC Press for their efforts.

Robert R. Luise

E-mail address:
hytem@voicenet.com

THE EDITOR

Robert R. Luise has spent a quarter of a century in research and development of high performance fibers and plastics materials, mainly at the DuPont Experimental Station, Wilmington, Delaware, from which he retired in 1993. He is the author or co-author of a number of important patents and publications, most notably in the field of liquid crystalline polymers. The latter include invention of a process for heat-strengthening of thermotropic fibers and other materials by post-polymerization and preparation of novel thermotropic advanced composite materials. He holds a Ph.D. degree in chemistry from the Massachusetts Institute of Technology and is currently President of Hytem Consultants, Inc., an independent consulting firm.

CONTRIBUTORS

B. C. Benicewicz
Polymers and Coatings Group
Los Alamos National Laboratory
Mail Stop E549
Los Alamos, New Mexico 87545
USA

P. N. Bier
Technology Manager
Polycarbonate Research
* and Development*
Bayer AG Plastics Business Group
Krefeld, Germany

A. L. Brody
Managing Director
Rubbright·Brody, Inc.
1285 Corporate Center Drive
Eagan, Minnesota 55121
USA

E. P. Douglas
Polymers and Coatings Group
Los Alamos National Laboratory
Mail Stop E549
Los Alamos, New Mexico 87545
USA

C. Feger
Research Division
IBM Corp.
T.J. Watson Research Center
Route 134
Yorktown Heights, New York 10598
USA

K. Friedrich
Professor and Technical Director
Institute for Composite Materials Ltd.
University of Kaiserslautern
Erwin-Schroedinger Strasse
67663 Kaiserslautern, Germany

E. M. Grossman
Vice President of Operations
SCORTEC, Inc.
2200 Renaissance Blvd.
Suite 250
Gulph Mills, Pennsylvania 19406
USA

R. P. Hjelm, Jr.
Manuel Lujan Neutron Scattering Center
Los Alamos National Laboratory
Mail Stop E549
Los Alamos, New Mexico 87545
USA

R. S. Irwin
Senior Research Fellow
DuPont Central Research
* Experimental Station*
E302/221C
P.O. Box 80302
Wilmington, Delaware 19880-0302
USA

D. A. Langlois
Polymers and Coatings Group
Los Alamos National Laboratory
Mail Stop E549
Los Alamos, New Mexico 87545
USA

R. R. Luise
President
Hytem Consultants, Inc.
1521 Winding Brook Run
Boothwyn, Pennsylvania 19061
USA

R. W. Lusignea
President
Superex Polymer, Inc.
350 Second Ave.
Waltham, Massachusetts 02154-1196
USA

W. G. Paul
Technical Marketing Manager
Polymers Division
Polycarbonate Resins
Bayer Corp. (formerly Miles, Inc.)
100 Bayer Road
Pittsburgh, Pennsylvania 15205-9741
USA

R. B. Prime
Senior Scientist
IBM Corp. Storage Products Division
5600 Cottle Road
San Jose, California 95193
USA

G. A. Serad
Technology Manager
Advanced Technology Group
Hoechst Celanese Corp.
2300 Archdale Dr.
Charlotte, North Carolina 28210-4500
USA

M. E. Smith
Polymers and Coatings Group
Los Alamos National Laboratory
Mail Stop E549
Los Alamos, New Mexico 87545
USA

C. E. Sroog
President
Polymer Consultants, Inc.
3227 Coachman Rd.
Wilmington, Delaware 19803
USA

G. Strupinsky
Vice President
Plastic and Packaging Research
Rubbright-Brody, Inc.
1285 Corporate Center Dr.
Eagan, Minnesota 55121
USA

D. J. Williams
President and Technical Director
The PST Group
P.O. Box 1543
Stowe, Vermont 05672
USA

CONTENTS

Preface iii

Contributors vii

I. LIQUID CRYSTALLINE POLYMERS

Chapter 1
Applications and Markets for Thermotropic Liquid Crystalline
Polyesters (TLCPs) 3
D. J. Williams

Chapter 2
Advances in Heat-Strengthened Thermotropic Liquid Crystalline
Polymer Materials 25
R. R. Luise

Chapter 3
Extrusion of Oriented LCP Film and Tubing 41
R. W. Lusignea

Chapter 4
SCORIM: Principles, Capabilities, and Applications to Engineering and
High Temperature Polymers 61
E. M. Grossman

Chapter 5
Properties of Liquid Crystalline Thermosets and Their Nanocomposites 79
E. P. Douglas, D. A. Langlois, M. E. Smith, R. P. Hjelm, Jr., and B. C. Benicewicz

II. POLYIMIDES AND ELECTRONICS

Chapter 6
Polyimide Synthesis and Industrial Applications 99
C. E. Sroog

Chapter 7
Electronic Applications of High Temperature Polymers 123
R. B. Prime and C. Feger

III. FIBERS

Chapter 8
High Temperature Fibers: Properties and Applications 149
R. S. Irwin

Chapter 9
PBI Fiber and Polymer Applications 161
G. A. Serad

IV. HEAT-RESISTANT MATERIALS

Chapter 10
Packaging Applications for High Temperature Plastics 181
G. Strupinsky and A. L. Brody

Chapter 11
New High-Heat Polycarbonates: Structure, Properties, and Applications 203
W. G. Paul and P. N. Bier

Chapter 12
Wear Performance of High Temperature Polymers and Their Composites.....221
K. Friedrich

INDEX 247

Part I

LIQUID CRYSTALLINE POLYMERS

Chapter **1**

APPLICATIONS AND MARKETS FOR THERMOTROPIC LIQUID CRYSTALLINE POLYESTERS (TLCPs)

David J. Williams

CONTENTS

I.   Introduction: Three Key Emerging Developments 4

II. The PST Group TLCP Growth Scenario 5
     A.   The Current Commercial Status of TLCPs 5
     B.   Forecast for the Next 5 Years 5
     C.   The Long-Range Outlook: Beyond the Year 2000 5
     D.   Developments Required to Fuel TLCP Growth 6

III. TLCPs Display Unique Combinations of Properties 6

IV. TLCP Compositions 7
     A.   Leading Commercial TLCP Compositions 7
     B.   What About NDC-Containing Compositions? 7

V.   Trends in TLCP Supplier Status 8
     A.   The Early Years 8
     B.   The Current U.S. Commercial Scene 8
     C.   The Changing Japanese Supplier Outlook 8
     D.   The European Scene 10

0-8493-7672-6/97/$0.00+$.50
© 1997 by CRC Press, Inc.

VI.   Profiles of Leading TLCP Players 10

VII.   Prices, Monomer Supply, and Prospects for Growth 12
       A.   Sample Price Calculations for an NDC-Based TLCP 13

VIII. Applications and Markets 15
       A.   Injection-Molded TLCP Parts 16
       B.   TLCP Films.. 17
             1.   TLCP Films in Electronic Applications 17
             2.   The PEN-NDC-TLCP Connection........................ 17
             3.   TLCP Film Barrier Properties 19
       C.   Vectran HS vs. Kevlar Fibers 20

IX.   Market Potential for Major TLCP Products........................ 21
       A.   Emerging Applications and Markets for TLCPs............. 21

X.   Summary and Conclusions.. 23

References 24

I. INTRODUCTION:
THREE KEY EMERGING DEVELOPMENTS

Thermotropic liquid crystalline polyesters (TLCPs) display remarkable and unique combinations of properties. Because of their relatively high prices, which are mainly attributable to high monomer costs, TLCP use has been limited, and TLCP markets have been slow to develop. TLCPs have been mainly used in advanced electronic devices, and increasingly in fiber optic components, where their properties are required and where cost-performance benefits justify their use. We believe that the three emerging developments outlined below and discussed in detail in this chapter signal an era of dramatic growth for TLCPs.

1. The Amoco Chemical 50,000 metric ton (t)/year dimethyl-naphthalene-dicarboxylate (NDC) plant, which came onstream in 1996, will provide significant downside pricing potential for TLCPs. NDC can be an effective thermotropic monomer in TLCP compositions. Priced at $2.00/lb, NDC will be less costly than other thermotropic monomers currently being used. [As noted below, this plant was built primarily for the manufacture of poly(ethylene naphthalate) (PEN).]

2. Hoechst Celanese and Polyplastics recently completed the construction of a 2800 t/year TLCP plant in Fuji City, Japan. This plant "responds" to a 30% growth rate in Asian demand for TLCPs, driven by the high concentration of electronic device producers in this area. At full capacity,

this plant will approximately double the global capacity of Hoechst Celanese and its TLCP partners to 6000 t/year. This capacity is not much less than today's estimated global TLCP consumption of 7000 t/year, and it is significantly larger than the capacity of any other Japanese supplier.

3. DuPont has emerged as a major global TLCP player with the recent completion of its full-scale, $20 million production plant. The capacity of this plant is said to be comparable to that of competitive TLCP plants. It is believed that DuPont's TLCP products, now marketed under the Zenite trade name, are based on NDC.

II. THE PST GROUP TLCP GROWTH SCENARIO

We first stated our TLCP growth scenario about 3 years ago. Our forecast seems to be on schedule. An update of our vision for TLCP commerce follows.

A. The Current Commercial Status of TLCPs

- Thermotropic liquid crystal polyesters (TLCPs) were first introduced into the commercial mainstream in the mid-1980s.
- TLCPs represent a young, emerging family of high-performance polymers, with an extraordinary combination of properties.
- TLCPs are expensive — from about $7.00/lb for injection-molding compounds to $22.00/lb for resin — mainly because of high monomer costs.
- Global consumption is low — currently about 7000 t/year.
- Electronic interconnect devices (EIDs) are the principal application.

B. Forecast for the Next 5 Years

- Growth in consumption, driven largely by electronics industry demand, ranges from 20 to 30%/year, depending on geographic region.
- Driven by lower monomer costs and higher competition, TLCP prices will fall to half their current levels during the next 5 years.
- By 2000, global TLCP resin demand will reach 18,000 to 23,000 t/year.

C. The Long-Range Outlook: Beyond the Year 2000

- We foresee the potential for TLCP resin prices as low as $2.00/lb — but not necessarily with today's leading compositions.

- Based on the price exclusion model, the potential for global consumption of TLCP resins will reach approximately 230,000 t/year, not counting the potential for fibers.

D. Developments Required to Fuel TLCP Growth

- Construction of world-scale monomer/resin production facilities to allow economies of scale and low resin prices is required: Amoco Chemical 50,000 t/year NDC plant is a major step in this direction; construction of a world-scale plant to produce p-hydroxybenzoic acid (HBA) would represent an even more significant step.
- Development of a broad array of applications for TLCP resins and compounds in diverse markets, ranging from electronics and fiber optics to aerospace, automotive, and food packaging, is needed.
- Development of appropriate TLCP compositions, blend technologies, processing equipment, and protocols enabling the economical, high-volume production of TLCP- based or TLCP-containing products using injection molding, extrusion (for fibers, films, sheets, tapes, and pro-files), blow molding, and thermoforming is necessary.

III. TLCPs DISPLAY UNIQUE COMBINATIONS OF PROPERTIES

Main-chain, all-aromatic TLCPs are high-performance, melt-processible thermoplastics with unique and remarkable combinations of properties. Their most prominent properties are

- Facile flow and processing
- Precision part fabrication
- Dimensional stability
- Low linear CTEs
- Combined stiffness, strength, and toughness
- Ceramic-like chemical resistance
- Unsurpassed permeation resistance
- Ultrahigh HDTs even without fiber reinforcement

Properties common to all-aromatic polyesters are:

- Low-moisture absorption
- Inherent flame retardance
- High-thermal stability
- High-radiation resistance

IV. TLCP COMPOSITIONS

All-aromatic, main-chain TLCPs comprise two types of backbone segments. This is illustrated in Figure 1, using representative compositions of two leading TLCP suppliers — Hoechst Celanese and Amoco Polymers.

Structure I: Hoechst Celanese Vectra

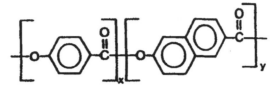

Structure II: Amoco Polymers Xydar

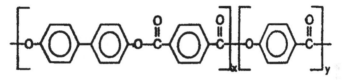

FIGURE 1
Two prominent families of commercial TLCPs.

A. Leading Commercial TLCP Compositions

- Rigid-rod, para-linked aromatic segments, such as the oxybenzoyl units in the illustrated compositions, impart liquid crystalline behavior. The principal mesogenic monomer in commercial TLCPs is 1,4-hydroxybenzoic acid (HBA). Generally, a 60 to 80 mol% para-aromatic moiety content is required to achieve liquid crystalline behavior.

- Non-para-aromatic segments, such as the non-oxybenzoyl units shown, impart thermotropic behavior by partially disrupting the rigid-rod structure imparted by para-linked units. We refer to the monomers responsible for this behavior as thermotropic monomers. As illustrated, the thermotropic monomers of today's leading TLCPs are 2,6-hydroxynaphthoic acid (HNA) and 4,4'-biphenol (BP) in combination with terephthalic acid (TA).

B. What About NDC-Containing Compositions?

- NDC is an effective thermotropic monomer that has not been widely available or widely used in commercial TLCP compositions.

- As emphasized in this presentation, the imminent availability of NDC at a relatively low price (for a thermotropic monomer) is a key development that will drive expanded TLCP sales.

V. TRENDS IN TLCP SUPPLIER STATUS

A. The Early Years

- TLCPs were introduced into the commercial mainstream in the mid-1980s with great fanfare by the Celanese Corp. (now Hoechst Celanese) and Dart Manufacturing Co. (now Amoco Polymers).

- The promise associated with early TLCPs prompted numerous other resin suppliers to announce TLCP development programs. By the late 1980s, these included DuPont, Granmont, Eastman, ICI, Bayer, BASF, DSM, Granmont, Rhône-Poulenc, Sumitomo, Polyplastics, and Unitika.

- This crowded situation and the realization that TLCP commerce would develop slowly prompted a full retreat by many of these "would be" players by 1990. Today's players are listed in Table 1, along with the compositions of their leading products.

- The U.S. and Europe now only have three suppliers — U.S.-based Hoechst Celanese, DuPont, and Amoco Polymers. Japan has 12 active players who serve the Japanese and Asian markets.

- The pattern of TLCP development in the U.S. and Europe is a classic example of the wave phenomenon. We believe that the industry is about to experience the onset of a second, larger, sustained wave.

B. The Current U.S. Commercial Scene

- The U.S. once clearly dominated the TLCP market, producing 90% of the world's TLCP resins. This commercial situation changed with the start-up of Polyplastics' plant, which uses Hoechst Celanese technology and with the emergence of Toray as another significant supplier. However, U.S. companies or U.S. subsidiaries of foreign-owned companies still hold the bulk of the world's TLCP-related patents.

- The world's leading suppliers — Hoechst Celanese, DuPont, and Amoco Polymers — are U.S. based. Hoechst Celanese has the largest share. For now, Amoco Polymers occupies second place. DuPont is challenging this order with the recent start-up of its full-scale production plant.

C. The Changing Japanese Supplier Outlook

- Table 1 identifies 12 Japanese companies engaged in TLCP commercial activity. The most prominent are Polyplastics (a joint venture between Hoechst Celanese and Daicel Chemical Industries), Toray, Sumitomo Chemical, and Unitika.

- Until recently, Japan accounted for about only 10% of the world's TLCP commerce. Japan's TLCP trade is restricted to Asia, where the heady

TABLE 1.

TLCP Product Suppliers

Leading World Players			
Company	*Trade Name(s)*	*IMC Types by HDT[a]*	*Composition of Leading Product*
U.S.			
Hoechst Celanese	Vectra and Vectran IMCs, films, fibers	Medium	HBA/HNA
DuPont	Zenite	Medium	HBA/NDC/HQ
Amoco Polymers	Xydar	Medium and high	HBA /BP/TA IA (optional)

Europe

Hoechst AG   Markets Hoechst Celanese products in Europe

Major Japanese Importers

Company	*U.S. Affiliate*	*Trade Name*	*30% Glass-Reinforced IMC Types by HDT[a]*	*Composition*
Polyplastics	Hoechst Celanese	Vectra	Medium	I IBA/HNA
Nippon Petrochemicals	Amoco Polymers	Xydar	Medium, high, and ultrahigh	HBA/BP/TA IA (optional)
Kuraray	Hoechst Celanese	Vectran	Fibers[b]	HBA/HNA

Japanese Suppliers Ranked by Capacity

Company	*Capacity t/year*	*Trade Name*	*IMC[a] Types by HDT*	*Composition*
Polyplastics	2800 ('96)	Vectra	Medium	HBA/HNA
Toray[c]	1000	Toray LCP	Medium	
Sumitomo Chemical	600	Sumika Super	Medium, high, and ultrahigh	HBA/BP/TA IA (optional)
Unitika	300	Rodrun	Low	HBA/PET
Ueno Fine Chemical Industry	300	Ueno LCP	Medium	HBA/HNA
Mitsubishi Kasei	200	Novaccurate	Low	HBA/PET
Mitsubishi Gas Chemical	50			HBA/NDC/HQ
Idemitsu Petrochemical	30	Idemitsu LCP	Medium	
Kawasaki Steel	30			HBA/NDC/HQ
Tosoh	30	Tosoh LCP	High	

[a] IMC, injection-molding compound; HDT, heat deflection temperature.

[b] Kuraray imports resin and spins Vectran fiber for Hoechst Celanese. Capacity estimated to be 500 t/year.

[c] Scheduled to come onstream in April 1997.

Source: The PST Group.

30%/year growth rate is largely driven by the high concentration of electronic manufacturing in that region.

- Polyplastics' 2800 mt/year TLCP plant, which came onstream in late 1995, will dramatically alter the Japanese supplier landscape. At full capacity, this plant will approximately double the capacity of Hoechst Celanese and its partners to 6000 t/year, and it will be significantly larger than that of any other Japanese supplier (Table 1). Will this plant cause smaller, less-qualified Japanese suppliers to exit the TLCP game?

- Toray Industries' recently announced plans to convert a polyester plant to produce 1000 t/year of TLCPs starting in April 1997 will add another significant player to the Asian TLCP market.

D. The European Scene

- Europe currently lacks a major "native" TLCP producer, although a number of European resin suppliers have unannounced development programs.

- Optatech (Espoo, Finland), founded in 1995 as a Neste Corporation spinoff, is developing a line of aliphatic-aromatic TLCPs, which are semi-compatible with polyethylene, for packaging applications.

VI. PROFILES OF LEADING TLCP PLAYERS

The Hoechst Celanese TLCP product line was launched by Celanese in 1985. Hoechst Celanese (formed when American Hoechst acquired Celanese in 1987) markets the broadest range of products in the TLCP business worldwide, under the Vectra (resins and injection-molding compounds) and Vectran (fibers) trade names. Hoechst Celanese is the only company to make TLCP-based films and fibers as well as resins and injection-molding compounds. Its position in Europe is strong due to the prominence of its parent, Hoechst, AG. It markets TLCPs in Japan through Polyplastics — a joint venture with Daicel Chemical Industries. In addition, it has a licensing arrangement with Kuraray to manufacture Vectran fibers (see below). We estimate that Hoechst Celanese currently enjoys about 60 to 70% of the global TLCP market.

Amoco Polymers acquired its Xydar TLCP business from Dart Manufacturing Co. in 1988. Dart acquired its technology from Carborundum in 1984 to secure a source of TLCP for their Tupperware™ line of dual-ovenable cookware. They established a 10,000 t/year facility in anticipation of broad acceptance of TLCP-based ovenware products. This acceptance never materialized, and now TLCPs are used for this purpose only in Europe and Asia. This application is believed to consume only 1000 t/year of Xydar SRT-30 0 — an ultra-high-heat TLCP. We estimate that Amoco has 20 to 30% of the world's TLCP market, including that sold for Tupperware production and Xydar injection-molding compounds.

DuPont began marketing DuPont LCP products in 1988, sans trade name, using pilot-scale facilities. DuPont has a strong tradition in LCPs, dating back to 1972 when they introduced their lyotropic LCP Kevlar™ fiber products. Early DuPont TLCP technology, dating from 1976, was based on phenyl hydroquinone (PHQ) with TA. With PHQ priced at about $18.00/lb and no downside pricing potential, early DuPont TLCP was too expensive to sustain commercial interest.

DuPont is now marketing non-PHQ-based (believed to be NDC-based) TLCPs, under the Zenite trade name. As noted above, DuPont began producing Zenite in a full-scale facility in late 1995. DuPont is focusing on molding compounds for electronic surface mount devices and encapsulation compounds for antilock brake sensors. The potential vulnerability of their Kapton™ polyimide film and Kevlar™ PPTA fiber businesses to displacement by TLCP-based films and fibers may be a motivating factor in their commitment to TLCPs.

Toray recently announced plans to convert a polyester plant to produce 1000 t/year of TLCPs. This will make them a significant player in Asia.

Polyplastics — a (45/55%) Hoechst Celanese/Daicel Chemical Industries joint venture and the largest Japanese TLCP supplier — compounds Vectra resins supplied by Hoechst Celanese. As outlined above, Polyplastics is destined to become a major force in TLCPs via its 2800 t/year plant. The Polyplastics plant will provide key economic advantages: it will be close to a major supplier of HBA and HNA monomer (Ueno, Table 3). Production in Japan will greatly reduce shipping costs and eliminate import/export duties in serving the Japanese market and in supplying Kuraray with resin for fibers. It will reduce the costs of serving the Asian market.

Kuraray manufactures Vectran fiber under license from Hoechst Celanese. Capacity is said to be 400 t/year. The Vectran price in Japan is said to be equivalent to that for Kevlar. One large-volume application under development in Japan is swordfish-resistant fishing nets. The new Polyplastics plant should also be a positive economic factor in Vectran development.

Sumitomo Chemical purchased the Carborundum TLCP technology — as did Dart Manufacturing — in 1972. Like Amoco products, the Sumitomo first products were based on the Carborundum original HBA/BP/TA chemistry. They are exporting the product to Taiwan. They recently brought a lower priced, medium heat TLCP to the marketplace.

Unitika acquired the original Eastman HBA/poly(ethylene terephthalate) (PET) low-heat TLCP technology and began marketing an improved version in 1987 under the Rodrun trade name. This improvement is attributed to a more random structure in the Unitika material which makes more efficient use of the HBA units.

VII. PRICES, MONOMER SUPPLY, AND PROSPECTS FOR GROWTH

For TLCP consumption to grow significantly, price reductions will be necessary. Some current list prices for TLCP products and their monomers are provided in Tables 2 and 3. The main factor regulating TLCP prices has been monomer prices. Currently, the only low-cost, readily available monomer suitable for TLCP production is TA (about $0.50/lb); unfortunately, there are no comparably priced para-aromatic diols, with which it can be coupled. Two monomer cost situations generally prevail: (1) monomer costs are high because of low-production volumes; and (2) in addition to low production volumes, costs are high because of inherently complex (and expensive) syntheses. These cases are illustrated below for HBA, HNA, and NDC.

A **first case example** — HBA (the favored mesogenic monomer in today's TLCP products) at about $2.40/lb is expensive only because of low-production volumes in batch facilities. Industry experts estimate that a continuous, world-scale plant with a 25,000 to 50,000 t/year capacity could readily produce HBA for $1.00/lb, with $0.50/lb being a long-term possibility. With today's leading TLCPs typically containing 60 to 80 mol% HBA, $1.00/lb HBA would dramatically impact their prices.

A **second case example** — HNA (the thermotropic monomer in the Hoechst Celanese Vectra) is not readily available, and pricing information is restricted. We estimate a current cost of about $8.00/lb. Because its synthesis is inherently complex (and expensive), HNA has limited downside pricing potential, barring a technical breakthrough. Will the emergence of NDC at $2.00/lb cause Hoechst Celanese to alter the Vectra composition?

An **emerging first case example** — NDC is an effective thermotropic monomer that is currently priced at about $5.50/lb. With the Amoco Chemical 50 t/year NDC plant, due to come onstream in 1995, the NDC price will fall to $1.50/lb for (relatively) high-volume PEN producers and $2.00/lb for low-volume TLCP producers. As the data in Table 3 show, NDC at $2.00/lb will be a relatively low-priced thermotropic monomer.

As the Amoco two-tier pricing suggests, the justification for its NDC plant is the potential for poly(ethylene naphthalate) (PEN), and not TLCPs. We believe that the NDC price for TLCP producers could fall to that for PEN producers as the consumption of NDC, PEN, and TLCP grow and other NDC producers emerge. Other industry sources have suggested that the NDC price could eventually fall to $1.00 as production volumes expand and other producers enter the business. Two key questions arise: will NDC become the thermotropic monomer of choice among TLCP suppliers and what about the patent situation?

TABLE 2.

Current TLCP Product Pricing

All-Aromatic Grades	Approximate List Price ($/lb)
Extrusion (unfilled)	12.00–22.00
Fibers	30.00–300.00
Films	44.00
Injection-molding compounds	
Mineral filled	5.25–10.35
Glass filled	7.15–10.65
Carbon filled	17.00–20.00

Source: Plastics Technology pricing update and industry interviews.

TABLE 3.

Some Representative Prices & Suppliers of Specialty TLCP Monomers

Current Price, $/lb	Low Price Potential, $/lb[a]	Supplier(s)	User(s)
HBA $2.50/lb	Excellent but no plans announced $1.00/lb.	Napp Chemicals (U.S.), Ueno Fine Chemicals (Japan)	Widespread
HNA $8.00/lb	Poor $8.00/lb	Ueno Fine Chemicals	Mostly Hoechst Celanese
BP $5.00-$7.00/lb	Good but no plans announced $2.00/lb	Schenectady (U.S.); Honshu Chemical and Mitsubishi Petrochemical (Japan)	Amoco, Sumitomo
NDC $5.50/lb	Excellent $1.50 for PEN $2.00 for TLCPs	Primarily Amoco Chemical; Mitsubishi Gas Chemical, Kawasaki Steel and Nippon Steel Chemical (Japan)	Limited now; expected to become widespread
HQ $2.50/lb	Modest $2.00/lb	Eastman Chemical Co.	Limited now; could grow with NDC use

[a] For volume production in continuous facilities.

Source: The PST Group.

A. Sample Price Calculations for an NDC-Based TLCP

As an exercise to gauge the effects of monomer costs and the state of market maturity on TLCP resin prices, we speculate about the possible history of an NDC-based TLCP. Our model material comprises 70/15/15 mol% HBA/NDC/HQ. We use the following monomer costs: $2.00/lb and $1.50 for NDC; $2.50 and $2.00/lb for hydroquinone (HQ); and $2.50

and $1.00/lb for HBA. The monomer costs per pound of our NDC-based TLCP resin for these pricing scenarios are shown in Table 4.

Reasonable markup ratios of conventional polyester resin prices over monomer costs for emerging and mature products are 3 and 2.5, respectively. For example, the 1993 and 1994 markup ratios for PET bottle resin (a mature material) were about 2.3 to 2.1, respectively. A higher ratio, such as 3.5, can be justified for emerging TLCP resins because of the higher costs associated with developing applications and processing technologies for these high-priced, low-volume, difficult-to-process materials. As shown in Table 4, we have used multiplier s of 3.5, 3.0, and 2.5 as measures of our TLCP market maturity.

Our most conservative scenario assumes a 3.5 markup ratio and no cost reductions for NDC (or HQ) as NDC, TLCP, and PEN sales volumes increase. Our most optimistic scenario assumes that the markup ratio will eventually reach 2.5 and monomer costs will fall to the lowest values indicated above and in Table 4.

The top row of Table 4 presents our most conservative NDC-based TLCP pricing scenario: (1) an emerging resin price of $10.50/lb, which is significantly less than the current $12.00 to $22.00/lb list prices of unfilled extrusion TLCP resins (Table 2); (2) a price falling toward $9.00/lb as the market grows; and (3) a resin price falling to about $7.50/lb as the market matures.

TABLE 4.

Estimated Prices for a 70/15/15 mol% HBA/NDC/HQ-Based TLCP

		Estimated resin price, $/lb[a]		
Assumed Monomer Cost		**Markup Over Monomer Cost**		
HBA/NDC/HQ Costs, $/lb	**Monomer Cost, $/lb of TLCP**	**3.5-Fold: Emerging TLCP**	**3-Fold: Growing Market**	**2.5-Fold: Maturing Market**
2.50/2.00/2.50	3.00	10.50	9.00	7.50
2.50/1.50/2.00	2.80	9.80	8.40	7.00
1.00/2.00/2.50	1.80	6.30	5.40	4.50
1.00/1.50/2.00	1.55	5.40	4.65	3.90

[a] neglecting any value of by-products, such as methanol.

Source: The PST Group.

If HBA remains at $2.50/lb (no world-scale plant constructed) and NDC and HQ costs decline to $1.50/lb and $2.00/lb, respectively, the top row pricing scenario just described will not change significantly. (The indicated changes in resin prices would be significant in a real pricing situation, but we do not consider them significant within the bounds of this exercise.)

At current HBA costs, further minor price reductions could be realized if a portion of the HBA were replaced with TA+ HQ. (Total HBA replacement is not possible because too high a TA/HQ content would promote too high a processing temperature.) With TA at $0.50/lb and HQ at $2.50/lb, on a *p*-phenyl moiety basis, the TA/HQ combination would represent about 60% of HBA cost.

If we assume that someone builds a world-scale HBA plant and sets HBA cost at $1.00/lb and that the industry matures sufficiently to justify a markup ratio of three, the price of our NDC-based model TLCPs would fall 40 to 45% into $4.65 to $5.40/lb range, depending on NDC and HQ costs. As the industry matures, our NDC-based TLCP price could fall below $4.00/lb.

The key points to be made include (1) with a 70 mol% HBA content, our model TLCP price is quite sensitive to HBA costs; this is generally true because the leading commercial TLCPs contain 60 to 80 mol% HBA. (2) Another factor is the Applications & Markets for TLCPs rate of industry maturation and its impact on the (resin price)/(monomer cost) ratio. This industry has been slow to mature, primarily because of high resin/monomer costs and secondarily because suitable processing and application technologies have been slow to develop. Thus, we have used conservative mark up ratios. (3) Key caveats are although we believe that we have outlined a reasonable scenario, the most difficult element to forecast is timing; another uncertainty is whether or when someone will build a world-scale HBA plant. These uncertainties are embodied in the phrase "beyond the year 2000" which we use in our summarizing statements. (4) Even without lower monomer costs, the price of our NDC-based TLCP could eventually fall to $7.50/lb as the industry matures, about half the current $12.00 to $22.00/lb list prices for unfilled extrusion resins. (5) The real breakthrough in HBA-based TLCP prices will occur with the construction of a world-scale HBA plant and its availability for $1.00/lb. By the time this occurs, the market should be mature enough to conservatively assume a markup ratio of three or less. Thus, with HBA priced at $1.00/lb, we estimate that the price of our NDC-based TLCP will fall to $4.50/lb. The best-case scenario estimates an eventual price of just under $4.00/lb.

VIII. APPLICATIONS AND MARKETS

Because of their current and near-term commercial importance, this chapter focuses on current and emerging applications and markets for:

- Injection molding compounds and molded precision parts, generally based on 30% glass fiber reinforcement
- Extruded films
- Spun fibers

A. Injection-Molded TLCP Parts

Of the products listed above, TLCP injection-molded parts have achieved the greatest commercial success. Their most prominent application is structural insulators in electronic interconnect devices (EIDs). EIDs intended for surface mount technology (SMT) infrared reflow solder (IR-RS) assembly is the fastest growing segment of this market and carries the most demanding performance standards. Growth rates for TLCP compounds in the EID arena range from 20 to 30%/year, depending on the region.

Table 5 compares the properties of the chief rivals in EIDs to TLCP compounds, using Hoechst Celanese Vectra E-130 as a representative example. These rival glass fiber reinforced compounds sell for $2.50 to $3.50/lb vs. $7.00 to $9.00/lb for TLCP compounds. The relatively steep prices of TLCPs are justified where their special attributes are required, such as in the manufacture and performance of long, thin-walled interconnect parts.

TABLE 5.

Comparison of Glass-Reinforced Infrared Solder-Tolerant Glass-Reinforced Injection-Molding Compounds

Material (Glass Fiber Content)	Vectra E130 TLCP (30%)	Tedur KU1-9510-1 PPS-II (40%)	Thermex CG-907 PCT (30%)	Amodel AF-1133 PAA (33%)[a]
Tensile strength, MPa	124	191	131	179
Elongation at break, %	2	3.9	1.8	—
Flexural modulus, GPa	13.7	14.9	10.3	13.1
Flexural strength, MPa	172	274.5	193	257
Izod impact strength, J/M				
Notched	107	101	82	80
Unnotched		668	507	
HDT @ 1.82 MPa, °C	270	265	260	273
Melting point, °C	355	285	289	310
CTE, ppm/°C, MD/TD	5/60[a]	22/28	22/—	20/45
Specific gravity	1.6	1.66	1.61	1.71
Mold shrinkage, %, MD/TD	0.1/0.2	0.1/0.5	0.25/0.7	—
Water absorbency, %	0.1[a]	0.004	0.06	0.18
Dielectric strength, kV/mm	15	14	16	18
Dielectric constant @1 kHz	3.7	3.6	3.7	4.2
Dissipation factor	0.06	0.002		0.15
	(1 mHz)	(1 kHz)		(1 mHz)
Volume resistance, ohm-cm	10^{15}	10^{16}	10^{15}	2×10^{16}
Arc resistance, sec	140	124	50	142
Price, $/lb	8.85	3.30	3.10	3.00

[a] Dry as molded.

Source: Supplier literature.

B. TLCP Films

We believe that films offer the second greatest market opportunity for TLCPs, after molded parts. As outlined below, we believe that TLCP films have the greatest near-term potential in electronic substrate applications and the greatest long-term potential in packaging applications.

To our knowledge, the sole commercial source of TLCP films is Hoechst Celanese, which began producing them in 1990. The principal difficulty in making TLCP films is controlling the high degree of machine-direction (MD) orientation that occurs under ordinary extrusion conditions. Hoechst Celanese and Superex Polymer (a Waltham, Massachusetts, plastics manufacturing and machinery company) are developing special machinery for this purpose. Films with biaxial orientation and balanced properties will be useful as electronic substrates and in barrier/packaging applications. Highly oriented films and tapes have potential in composite structures. Some properties of representative balanced films with which balanced TLCP films are expected to compete are compared in Table 6.

1. TLCP Films in Electronic Applications

TLCP films and the associated technology are being developed to compete with polyimide (PI) films in electronic applications. Vectra A 900 film is compared with Ube Industries (Japan) Upilex-S — a rigid-rod PI with outstanding performance credentials — in Table 6. Both films display similar tensile and CTE values — attributable to their rigid rod structures. Vectra A900 film, however, displays far less moisture sensitivity. Both have excellent dry electrical properties, so that the difference in moisture sensitivity is a factor favoring TLCP films in electronic applications. High moisture sensitivity also adversely impacts dimensional stability and promotes conductor corrosion. PIs do offer superior thermal stability; for example, useful service at 270 to 290°C has been reported for Upilex-S. The price advantage of TLCP films over PI films ($44.00/lb vs. $60.00/lb — Table 6) coupled with their vastly superior moisture resistance makes this a very logical, high value-added market entry point for TLCP films.

2. The PEN-NDC-TLCP Connection

PEN film is an upscale analog of PET film. Processed like PET, PEN films are clear and offer superior barrier performance, heat and chemical resistance, mechanical strength, and dimensional stability compared to PET. Potential applications of PEN include all those where PET might be used if it had PEN's properties. Potential electrical and electronic applications for PEN films include high temperature motor insulation, magnetic tapes, flexible circuits, and capacitors. Other potential uses are blow-molded bottles and microwavable and medical packaging. The huge, shelf stable

TABLE 6.

Properties of Representative Balanced Films

Company	ICI	ICI	Hoechst Celanese	Ube Industries
Film type	BO-PET[a] MD/TD	Kaladex PEN	Vectra TLCP	Upilex-S PI
Mechanical				
Tensile strength, MPa	165/228	220	345	413
Tensile modulus, GPa	4.1/4.8	6.0	8.3	8.9
Thermal and Physical				
T_g/T_m, °C	72/269	120/262	125/280	No/no
Upper use, °C[b]	130	150	180	260
CTE, ppm/°C	23/18	20	−5−+9	2.6
Moisture absorption, %	< 0.1		0.02	1.0
Chem/sol resistance	Good	Good	Excellent	Excellent
Rel oxygen permeation	100	25	1	
Electrical				
Dielec str @ 1 mil, V/mil	4500		5500	6800
Dielec constant, 1 KHz	3.2		3.3	3.5
Dissipation factor, 1 KHz	0.005		0.003	0.0013
Vol resistivity, ohm-cm	10^{18}		10^{16}	10^{17}
Cost				
Film price, $/lb	2–6	19–10	44	60

[a] BO-PET = biaxially oriented-PET
[b] Estimates only. Actually, upper service temperature depends on conditions of use and film treatment.

Source: Supplier literature.

food market — now largely inaccessible to PET because of temperature and barrier limitations — is an especially attractive opportunity for PEN.

Global commercial interest in PEN is developing rapidly. Teijin and ICI introduced the first commercial films made from PEN in 1989 and 1992, respectively, priced at about $18.00/lb. The commercial potential of PEN is the motivating factor behind the Amoco Chemical decision to build a 50,000 t/year NDC plant. NDC's price dropped from $5.50 to $1.50/lb (for large-volume users) when this facility came onstream in late 1995. This price reduction should translate to a PEN film price of about $10.00/lb. PEN resin is available from Eastman Chemical and ICI. Other major players interested in NDC and PEN include DuPont, NKK, Mitsubishi Gas Chemical, and Kawasaki Steel.

As indicated above, the development of NDC and PEN are extremely important for TLCPs. NDC is an effective thermotropic monomer, and the availability of NDC at $2.00/lb (for low-volume sales) is expected to reduce TLCP prices and promote their expanded use. As listed in Table 1, some commercial TLCP compositions are believed to be NDC based.

3. TLCP Film Barrier Properties

One of the most important and potentially useful properties of TLCP films is their unexcelled barrier performance. Figure 2 compares the water vapor and oxygen permeability of 1-mil TLCP film with those of several other polymeric films. The TLCP film is clearly superior in its ability to exclude both water and oxygen. It is believed that TLCP film is generally superior to all other thermoplastic films in its combined permeation resistance to water and oxygen.

It is often said that today's TLCPs are too expensive to be considered for most packaging applications, especially food packaging. This is simply not true. For example: TLCPs cost about 20 times more than PET. However, TLCP films are about 100 times more resistant to oxygen and water permeation than PET films (Figure 2). Hence, based on resin price only, TLCPs are about five times more cost effective. However, cost-effective processing technologies have not yet been developed to make TLCPs broadly competitive in packaging. The key point is that high resin cost per se is not the barrier keeping TLCPs out of packaging markets.

Military packaging, where long shelf life is required, provides a logical high value-added entry opportunity for TLCPs into the packaging market.

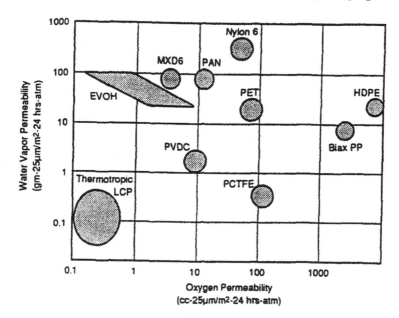

FIGURE 2
Permeability of various polymer films. (Reproduced with permission of Superex Polymer.)

C. Vectran HS vs. Kevlar Fibers

Because their prices are expected to remain high for the foreseeable future compared to competitive fibers, TLCP fibers will be the last of the major TLCP forms to gain a strong commercial position. Hoechst Celanese, with Kuraray, supplies Vectran HS — a heat-treated high-strength TLCP fiber. Its chief competitors are various lyotropic p-aramid fibers, especially poly(p-phenylene terephthalamide) (PPTA). The world leader in PPTA fibers is DuPont, with its well-known Kevlar trade name. Other LCP aramid fiber producers are Teijin Ltd. and Akzo Nobel.

Table 7 compares key properties of Vectran HS to three grades of DuPont Kevlar fibers. Its properties closely resemble those of Kevlar 49, and these fibers are therefore expected to compete for many of the same applications. Thus, commercial information on PPTA fibers can be used to gauge the potential for HS-TLCP fiber commerce.

Vectran HS fiber prices range from $30.00 to $300.00/lb. Kevlar fiber prices vary from $10.00 to $100.00/lb. An average price for 1500 denier

TABLE 7.

Key Property Comparisons: Vectran HS vs. Kevlar Reinforcing Yarns

Property	Vectran HS	Kevlar 29	Kevlar 49	Kevlar 149
Mechanical				
Tenacity, GPa	2.5–3.1	2.9	2.9	2.3
Elongation, %	2.2–2.5	3.6	2.5	1.5
Initial modulus, GPa	84–103	67	114	145
Percent creep				
After 10,000 h	0	3	0.4	NA
@ % break load	33%	50	50	NA
Thermal				
Axial CTE, ppm/°C	–5	–2.3	–4	–2
Radial CTE		67	63	67
Long-term use, °C		160	160	160
Use range: upper, °C	100	200	200	200
Use range: lower, °C	–200	–200	–200	–200
Physical				
Density	1.4	1.44	1.44	1.47
Dielectric constant	3.2	4.5	4.1	4.1
Moisture regain, %	< 0.1	4.3	4.3	1.5

Source: Hoechst Celanese and DuPont product literature National Materials Advisory Board, Commission of Engineering and Technical Systems, National Research Council (NMAB-453) National Academy Press, 1990.

Kevlar 29 is $15.00/lb. It seems reasonable that TLCP fibers will have to sell in the neighborhood of $15.00/lb to have a market potential equivalent to that of PPTA fibers. We estimate a current annual worldwide PPTA fiber market on the order of 50,000 t. We believe that the potential market for HS-TLCP fibers — at Kevlar-equivalent prices — is of the same order.

IX. MARKET POTENTIAL FOR MAJOR TLCP PRODUCTS

As stated above, we estimate the global market for TLCP resins at about 7000 t/year. This is attributable mostly to TLCP molding compounds (about 6,000 mt going to EIDs and 750 to 1,000 t going to miscellaneous precision molded parts). TLCP fibers probably enjoy a market volume of just over 500 t, and TLCP films probably just under 500 t. The annual growth rate is reported to range from 20% in the U.S. and Europe to 30% in Asia, where activity in electronic device manufacture is greatest. Table 8 provides a summary of some applications and markets for TLCP products. Some specific examples of emerging applications and markets are briefly described below.

We estimate current global TLCP capacity to be 15,000 t/year, including the Polyplastics and DuPont new plant capacities. This seems large compared to current demand of about 7000 t/year. With a 20 to 30% annual growth rate, however, it will only take a few years to reach full utilization of today's global TLCP capacity: 4.3 years at 20%/year and 3.0 years at 30%/year.

A. Emerging Applications and Markets for TLCPs

- Fiber optics: These connector components include bodies and ferrules, splice components and bodies, undersea coupler housings, diode receptacles, and Vectran fibers as strength members for flexible, coiled interconnect cables. Injection-molded connectors, splices, and coupler housings offer a direct challenge to machined metal and ceramic components, which now dominate this application.

- Computer cases: It is believed that several Japanese manufacturers of notebook computers are preparing to use TLCP molding compounds for the computer housings and portable telephones. Such use would constitute a major breakthrough in per-part TLCP compound consumption, a situation that could have a significant impact on overall TLCP compound consumption. We believe this use is based on the need to place circuitry on the housings to save space and fabrication costs. Such use apparently requires a "circuitizable" TLCP and surface mount technology (SMT) coupled with infrared reflow solder (IR-RS) assembly.

TABLE 8.

Markets Benefiting from TLCPs — Examples of Applications and Key Properties

Market	Application	Key Properties	Status
Electronics	Electronic interconnect devices (EIDs)	Molding precision Dimensional stability Solder resistance	Established and growing
	Flexible circuits	Low moisture absorbency Dimensional stability Solder tolerance	Emerging
Food	Packaging	Low permeability Sterilization compatible Hot-fill capability	Potential
Fiber optics	Connectors	Precision molding Dimensional stability	Emerging˙
	Cables strength members	Moisture resistance Dimensional stability	Emerging
Medical	Appliances, devices, tubing	Sterilization compatible Nonconductive	Emerging
	Biofluid and pharmaceutical packaging	Ultralow permeation Biocompatibility Sterilization compatible	Emerging
Automotive	Molded underhood components	Thermomechanical stability Chemical resistance	Emerging
	Composite fibers	High-specific strength High-specific modulus Near zero creep	Potential
Aerospace	Engine, electrical, structural components	High-specific strength High-specific modulus Low flammability Low smoke and toxic fumes	Emerging
Cryogenic	Pipes, tubes, and fluid transfer devices	Low permeability Excellent cryogenic properties	Emerging
Chemical	Reactor and storage tank liners	Low permeability Chemical resistance	Emerging
Commercial aircraft	Structural composites	Dimensional stability High strength/stiffness Thermal resistance	Potential

Source: The PST Group.

- Electronic device encapsulation: The encapsulation of electronic devices offers additional application opportunities. Examples include devices with stiff leads such as capacitors, thermistors, automotive underhood and ABS components, and undersea optoelectronic devices.

- Medical tubing: Superex Polymer has developed biaxially oriented extruded LCP tubing for minimally invasive laproscopic surgical instruments. Superex Tubing is designed to displace plastic-coated metal tubing now used in this application based on its excellent insulative and mechanical properties and lower cost.

- Speakers: Mitsubishi Electric has described the use of TLCP-based injection-molding compounds to manufacture speaker diaphragms.

- Auto bodies: Mazda is developing a TLCP-thermoplastic blend composite for the manufacture of automotive body panels

- Camera frames: Nikon is molding frames from TLCP compounds.

- Flexible circuits: TLCP films are expected to enter the marketplace soon via the electronic film and substrate market, by capturing share from PI films. We believe flexible circuits offer a major opportunity for market ingress by balanced all-aromatic TLCP films. A substantial opportunity for flexible circuits appears to be emerging in the automotive market, where there is a need for a cost-effective substrate material for under-hood flexible circuits. With a global 50 million unit per year production rate (includes cars, vans, and light-duty trucks), this represents a larger potential market for TLCP films than today's estimated 1500 t/year market for PI films.

X. SUMMARY AND CONCLUSIONS

- Main-chain, thermotropic liquid crystalline polyesters (TLCPs) are a significant new class of high-performance, melt-processable thermoplastics with unique and remarkable combinations of properties.

- TLCP cost-performance benefits have been well proven in precision injection molded parts — most notably in advanced electronic interconnect devices.

- TLCPs will soon challenge high-performance polyimide films in flexible electronic circuit applications.

- Applications and markets for TLCPs have been slow to develop because of high resin prices stemming from high monomer costs.

- Construction of a world-scale HBA plant and HBA availability for about $1.00/lb would allow a major reduction in TLCP resin prices.

- We have forecast TLCPs prices at about half their current levels and resin consumption of 18,000 to 23,000 t/year by the year 2000.

- Beyond 2000, we expect TLCP resin prices to be as low as $2.00/lb (but not necessarily with today's compositions), and consumption potential to climb toward 230,000 t/year (based on the price exclusion model), not counting applications for fibers.

- Three recent events — the start-up of Amoco Chemical NDC and Poly-plastics and DuPont TLCP plants — signal an era of expanded growth for TLCPs. They also provide evidence that our TLCP growth scenario is on schedule.

REFERENCES

This chapter is based in part on the PST Group 1992 multiclient study: *Technical and Commercial Issues Governing the Development of Thermotropic Liquid Crystal Polyesters in the 1990s.*

Also see Kirsch, M.A. and Williams, D.J. Understanding the thermoplastic polyester business, *ChemTech*, April 1994, 40.

Chapter **2**

ADVANCES IN HEAT-STRENGTHENED THERMOTROPIC LIQUID CRYSTALLINE POLYMER MATERIALS

Robert R. Luise

CONTENTS

I.   Introduction 26

II.   Historical Background 27

III. Heat-Strengthening 29
      A.   Fibers 29
           1.   Tenacity Model 30
           2.   Blockiness/Crystallization 30
           3.   Vectran Fiber 32
      B.   Films 32
           1.   Biaxial Films 32
           2.   Barrier Properties 33
      C.   Uniaxial Sheets 33
      D.   Pultruded Shapes 35
      E.   Molded Articles 35
           1.   Mechanical Anisotropy 37
      F.   Other Materials 37

0-8493-7672-6/97/$0.00+$.50
© 1997 by CRC Press, Inc.

IV.   Summary 37

References 38

I. INTRODUCTION

Main-chain thermotropic liquid crystalline polymers (TLCPs) represent a class of high performance condensation polymers composed mainly of para-extended aromatic polyesters.[1] Invented in U.S. industrial research laboratories in the early 1970s, development was inhibited until the 1980s in part by high monomer costs (principally of aromatic diols) and a time lag in ideal market opportunities. With improved prospects of lower cost comonomer availability, particularly of p-hydroxybenzoic acid and 2,6-naphthoic acid, the latter afforded by poly(ethylene naphthalate) (PEN) expansion, and rapidly expanding TLCP applications in electronics packaging, the TLCP market presently estimated at 10 million pounds now seems secure and ripe for 25% annual growth projected for the remaining 1990s.[2,3] Major commercial players include Hoechst Celanese with Vectra, Amoco with Xydar, and DuPont with Zenite, all aromatic copolyesters.

Often-quoted advantages of TLCPs include high thermal, electrical, and solvent resistance combined with excellent (unparalleled) orienting flow characteristics and dimensional stability — ideal for injection-molded applications such as thin-walled electrical insulators with close tolerances. In film form, excellent barrier properties are afforded by the dense uniaxial LC morphology,[4] and it is anticipated that medical and food packaging opportunities in addition to electronics will emerge rapidly coincident with lower resin costs and expanded TLCP production.

Perhaps the outstanding TLCP characteristic, certainly the one which attracted major industrial attention to these materials in the 1970s, is their ability in oriented fiber form to dramatically strengthen by post-heat treatment due to solid-phase polymerization.[5] The latter afforded unprecedented tensile properties for melt-spun polyesters, seen only before in Kevlar solution-processed polyamide fibers.

While main-chain TLCPs have been covered extensively in a number of reviews,[6-9] the historical perspective on these materials remains somewhat clouded, largely because essentially all of it lies in the often-overlooked patent literature. Moreover, much of the work related to post-heat strengthening of these materials, particularly in nonfiber form, remains embedded in the patent literature. The present chapter then, seeks to further clarify the historical perspective on these materials based on the published literature; and is also intended as a review of the unique poststrengthening attributes of these materials observed not only in fibers but also in other forms such as films, sheets, and pultruded and molded shapes.

II. HISTORICAL BACKGROUND

Based on the published literature, the earliest recognition of thermo-tropicity in aromatic copolyesters appears to be a German patent application to DuPont published on November 27, 1975,[10] U.S. filing May 10, 1974. This application disclosed liquid crystalline (or optically anisotropic) melts of *para*-aromatic copolyesters, mainly of the AA-BB type based on substituted hydroquinones, and also high-tenacity fibers and films obtained by solid-phase polymerization in the relaxed state (heat strengthening). It formed the basis for two Belgian patents granted November 12, 1975;[10] and four subsequent U.S. patents granted to each of the four original authors, including three compositional cases[11-13] and one process case for fiber heat-strengthening.[14]

While prior patents were granted to others, notably to Eastman[15] and Carborundum[16] on copolyesters based on p-hydroxybenzoic acid later shown to be thermotropic, there was no published recognition of liquid crystallinity in these materials, it appears, prior to the earliest DuPont publication date. For example, the first journal publication recognizing liquid crystallinity in poly(ethylene terephthalate) (PET)/p-hydroxybenzoic acid partially aromatic copolyesters patented in 1973[15] appeared in August 1976 (submitted in September 1975).[17] An earlier publication in February 1975[18] referred to the latter copolyesters as highly fluid and self-reinforcing due to unusually high (anisotropic) mechanical properties observed in the flow direction for injection moldings, but there was no mention of liquid crystallinity until the 1976 article. Published acknowledgement of liquid crystallinity in Carborundum polyesters based on 4,4'-biphenol, terephthalic acid, and p-hydroxybenzoic acid patented in 1976[20] did not appear until a journal article in April 1979.[27] Earlier fusible aromatic copolyesters patented by Carborundum in 1972 were based on hydroquinone, isophthalic acid, and p-hydroxybenzoic acid; and while not recognized as such then, were also found later to be thermotropic.[36] Hoechst–Celanese (earlier Celanese), the leader in TLCP commercial development, obtained its first TLCP patent in 1978 on liquid crystalline aromatic copolyesters based on p-hydroxybenzoic acid, 2,6-naphthoic acid, and terephthalic acid.[19]

As for high strength fibers, the DuPont disclosure[10,14] appears to be the first to recognize that fiber tenacity of condensation TLCPs, including the aromatic copolyesters and polyazomethines disclosed, could be substantially increased to Kevlar-like levels (~2.4 GPa or 20 gpd) by a post-polymerization of the fiber near the melting point in a purging inert atmosphere and in a relaxed or unconstrained state without decreasing spun-in orientation. Purging was required to remove by-products of solid-phase polymerization.[5] A Carborundum patent issued in 1976[20] also disclosed a process for high modulus copolyester fibers based on 4,4'-biphenol, terephthalic acid, and p-hydroxybenzoic acid by heating in an oxidizing atmosphere (e.g., air) followed by stretching (10 to 400%). Strength improvement (~2x to 1.4 GPa shown) was also observed combined with higher modulus and decreased elongation, typical of drawing processes, and was attributed to crystallization and cross-linking. Drawing was required for strength improvement. In contrast, significantly increased elongation and either unchanged or increased modulus accompanying greatly enhanced strength (generally four to five times) were observed by the DuPont process, without tension or drawing, consistent with a large linear molecular weight increase.

Although fibers of Eastman PET/p-hydroxybenzoic acid copolyesters were spun,[15] there is no evidence of high tenacity, presumably because these compositions are known to be biphasic (i.e., contain both liquid crystalline and isotropic phases), which inhibits heat strengthenability due to deorientation (or lack of spun-in orientation) of the isotropic component.[21] Celanese did not disclose high-tenacity TLCP fibers until issuance of a composition patent in 1979 based on Vectra-type copolyesters of p-hydroxybenzoic acid and 2,6-hydroxynaphthoic acid.[22] High strength was obtained by a heat treatment process which appears to be identical to that disclosed earlier by DuPont.[10] Kuraray, the Japanese co-producer of Hoechst–Celanese Vectra-based TLCP fibers (known as Vectran) has recently obtained a U.S. patent on a heat treatment process for high-strength TLCP fibers which appears to be similar to that of DuPont,[14] except that dry air is introduced in the latter stages of the heat strengthening, with improved abrasion resistance shown.[23]

Table 1 lists a chronology of important published TLCP events as described above.

TABLE 1.

TLCP Published Chronology

Year	Event	Ref.
1972	U.S. Patent (Carborundum), fusible aromatic p-oxybenzoyl copolyesters	16
1973	U.S. Patent (Eastman), PET-modified p-oxybenzoyl copolyesters	15
1975	German Patent Application (DuPont), liquid crystalline melts of p-aromatic copolyesters and high-tenacity fibers by solid-phase polymerization; two Belgian patents granted	10
1976	U.S. Patent (Carborundum), process for high-modulus p-oxybenzoyl aromatic copolyester fibers of improved strength by heating and stretching	20
1976	Journal article, discloses liquid crystallinity in Eastman PET-modified copolyesters patented in 1973	17
1976, 1978	U.S. Patents (DuPont), liquid crystalline copolyesters disclosed in 1975	11–13
1978	U.S. Patent (Celanese), liquid crystalline p-aromatic copolyesters	19
1979	Journal article, discloses liquid crystallinity in Carborundum p-oxybenzoyl copolyesters	27
1980	U.S. Patent (DuPont), process for high-tenacity fibers of TLCPs by solid-phase polymerization disclosed in 1975	14

III. HEAT-STRENGTHENING

As shown in early DuPont disclosures,[10-14] heat-strengthening consists of solid-phase polymerization principally of axially oriented as-spun fibers in a continuously purged inert atmosphere. Purging is required to remove post-polymerization by-products and drive the equilibrium polycondensation reaction forward.[5] The atmosphere should be inert or in vacuo to minimize undesirable side reactions which may cause branching (e.g., Fries rearrangement in polyesters). Preferred temperatures are near but below the melting point and heating stagewise since the melting point may increase as much as 30°C in the process. It is important not to exceed the melting point in heating, to prevent deorientation. Heating times are long (minutes-hours) because of diffusion limitations of condensation by-products. Fibers are preferably heated in the relaxed or free state, since constraints or loads result in breakage or excessive interfilament fusion and are detrimental to tenacity development. Orientation is critical for strength enhancement, and retention of it despite heating in the unconstrained state is a novel feature of the rigid extended chain uniaxial morphology which imparts exceptional dimensional stability (little or no change in fiber length observed).[5]

While heat-strengthening has been applied mainly to fibers, often overlooked are other TLCP materials discussed below which also respond to post-treatment such as films, uniaxial sheets, molded or extruded shapes, pultruded shapes, foams, pulps, and pressed fiber webs.

Table 2 summarizes the essential elements of heat-strengthening, including process, property, and structural changes.

A. Fibers

As noted above, fiber heat-strengthening has been covered most extensively in a number of reviews and research investigations.[6-9] The early DuPont disclosures[10-14] describe the process as outlined briefly above, including the use of inert coatings (e.g., talc or graphite) to reduce interfilament sticking and packaging options for yarns such as skeins,

TABLE 2.

TLCP Heat-Strengthening Essentials

Composition — Condensation TLCPs (e.g., aromatic copolyesters)

Precursor form — Preferably oriented: fibers, films, uniaxial sheets, foams, pultruded or extruded shapes, injection-molded articles

Process — Prolonged (minutes-hours) heat treatment to near but below the melting point in an inert atmosphere with continuous purge and little or no constraints or loads

Response — Increased tensile strength, elongation-to-break, toughness, thermal and solvent resistance, other molecular weight-dependent properties

Structural changes — Increased molecular weight, glass temperature, crystalline melting point and melting endotherm, retained orientation

---

soft-covered bobbins, and piling in perforated baskets.[10] Other reported investigations include studies of the effects of spinning variables and structure on heat treatment response by Krigbaum and co-workers[24,25] and a tenacity model proposed by Yoon[28] based on uniaxial chain-length growth.[28] Krigbaum et al. examined the effect of heat treatment response of Vectra-type TLCPs, documenting the importance of spin-stretch and as-spun orientation for tenacity improvement. They also observed increases in molecular weight, crystallinity, and crystalline melting point on heat treatment, consistent with other investigators.[26] Elongation increases, consistent with molecular weight growth, were also reported.[24]

1. Tenacity Model

Yoon's[28] uniaxial tenacity model[28] corroborated experimental observations[5,7,10] that tenacity increase on heat treatment is primarily due to an increase in chain length. Excellent fit of the model with experimental data was obtained assuming a most probable molecular weight distribution and a single shear strength for as-spun and heat-treated fibers, implying no change in interchain interaction (i.e., spacing) on heat treatment. Figure 1 is a plot taken from Reference 28 of tenacity vs. I.V. growth for as-spun and heat-treated fibers of HBA/HNA 73/27 Vectra (p-hydroxybenzoic acid[HBA], 2,6-hydroxynaphthoic acid[HNA]). Maximum as-spun tenacity of about 1.8 GPa appears to be controlled by viscosity/molecular weight spinning limits (I.V. ~7 corresponding to DP ~300 and mol wt ~20,000 from Figure 5 of Reference 28). Heat treatment affords molecular weight growth beyond apparent practical processing limits to I.V. ~30 (DP ~1000 and mol wt ~135,000 from Figure 5) and tenacity in the 3.5-GPa range. The latter is consistent with the tenacity/molecular weight relationship found for other TLCP fibers.[5]

Since strength conversion is tied to molecular weight growth, it follows that polymerization catalysts should be feasible to accelerate the strengthening process. Coatings of alkali salts afford such an advantage as described in a patent to DuPont,[30] whereby ~2x reductions in conversion times are shown with 0.015 to 2% aqueous solutions applied to TLCP copolyester yarns. The effectiveness of topically applied catalysts is perhaps not surprising for such surface-intensive materials. A Celanese patent[31] also describes similar accelerative effects on fiber heat-strengthening for alkali salts applied during polymerization prior to spinning. The importance of balanced end groups in the post-polymerization (transesterification) process is covered in another Celanese patent application, in which excess carboxylic acid end groups are used to suppress postreactivity for processing advantages.[32]

2. Blockiness/Crystallization

Because of the ready availability of Vectra over the years, there have been many structural and mechanistic studies of the HBA/HNA

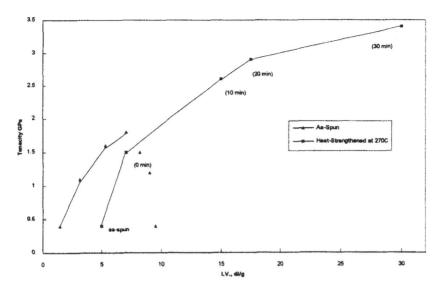

FIGURE 1
Tenacity/molecular weight relationship for HBA/HNA 73/27 fiber. (From Yoon, H.-N., Strength of fibers from wholly aromatic polyesters, Figures 1 and 2, *Colloid Polym. Sci.*, 268, 230, 1990. With permission.)

system — more than any other TLCP — and thus this composition has served as a useful model of TLCP behavior. While there seems to be a consensus that the copolymer as prepared by melt acetolysis consists of randomly distributed comonomers, there is less agreement as to the direction of blockiness after heat treatment in which crystallization and melting point increases are observed. For example, Muhlebach et al.[33] found a random arrangement of HBA and HNA comonomers by NMR sequencing,[33] as did Blackwell et al.[34] in X-ray fiber patterns of crystalline regions.[34] On heat treatment of filaments, Schmach and Vogel[35] interpret the appearance of a new DSC melting peak at 345°C as evidence of transesterification to a block copolymer,[35] but there is disagreement between the DSC and wide-angle X-ray evidence.[26] The latter shows a hexagonal to orthorhombic lattice transition on heat treatment, but no change in comonomer sequencing.[68] The latter was also found by Blackwell et al. in X-ray analysis of as-spun and heat-treated fibers based on 4,4'-biphenol, terephthalic acid, and *p*-hydroxybenzoic acid.[55] X-ray shows, in general, a sharpening of lateral ordering in heat strengthening of TLCP fibers with no major effect on longitudinal order, the presence of which does not appear to affect the strengthening process.[5] Rather, it is the lateral order (i.e., axial orientation) which must at least be retained to allow strengthening. As for DSC melting peaks, the melting point increases appear to be related to the postpolymerization process, but are (unlike X-ray) orientation-insensitive.[29] Hence, the combination of high lateral order (X-ray) and increased melting point (DSC) is probably a necessary and sufficient indicator of heat strengthening provided no deoriented amorphous phases are present.[5] The latter should be detectable by dynamic mechanical analysis (DMA) or birefringence measurements.

3. Vectran Fiber

Presently, the only commercial TLCP fiber is the Hoechst–Celanese Vectran, introduced in 1988. The fiber is based on the Vectra HBA/HNA 73/27 copolyester, and is produced in both as-spun and heat-strengthened forms, the latter by Kuraray in Japan. Basic fiber properties of heat-strengthened Vectran HT are similar to those of Kevlar 29 aramid, as shown in Table 3.[57] A high-modulus version, Vectran UM, is also available with properties similar to high-modulus Kevlar 49, presumably for composites uses. Noted advantages of Vectran HT vs. Kevlar include superior abrasion and creep resistance, acid/alkali resistance, moisture absorption, and hot-wet resistance (primarily for rope applications); however, high cost ($20 to $40/lb) is an impediment to broad market development, and high temperature properties should be inferior to Kevlar.

TABLE 3.

Basic Properties of Vectran HT and Kevlar 29 Fibers

	Vectran HT	Kevlar 29
Density, g/cc	1.41	1.44
Melting point, °C	330	>600
Tensile strength, GPa	3.2	2.7
Elongation-to-break, %	3.8	3.9
Tensile modulus, GPa	72.9	69.3
Limiting oxygen index, %	28	30

B. Films

Heat-strengthened TLCP film was first disclosed in the 1975 DuPont patent application.[10] A U.S. patent to Celanese in 1982[37] claims improved flow orientation and mechanical properties for Vectra-type extruded film using a grid of cone-shaped holes in the die. Improved orientation with the grid resulted in a larger heat-strengthening response (two times vs. starting materials) with MD tensile strengths to 0.83 GPa. This compares with a much greater tenacity response (up to ten times) observed for more highly oriented fibers.[14,28]

1. Biaxial Films

Since high MD film orientation leads to poor transverse (TD) and tear mechanical properties, blown film techniques have been utilized to improve the property balance. A European patent application to Kuraray[38] uses conventional film blowing at high shear rates (800 to 1200 s⁻¹) to obtain balanced Vectra films, while Superex uses a counter-rotating die extrusion process to obtain biaxial Vectra films.[39] Another route to a biaxial film is afforded by lateral stretching of uniaxial sheets prepared from warps of TLCP fibers, as discussed further below.[5,47] Balanced film tensile strengths in the range of 100 to 250 MPa can be produced by this method, similar to extruded biaxial films.[44] In still another route, cholesteric TLCP films with improved property balance by conventional processing have been reported in U.S. patents to Nippon Oil.[40–43] The chiral state is obtained by TLCP modification with chiral intermediates such as 3-methyl adipic acid[40] and 3-methyl-1,4-butanediol.[42] Under certain conditions, transparent monodomain films with ordered planar texture were produced.[41] There is little information on the heat-strengthening response of these materials.

2. Barrier Properties

Excellent permeability resistance to gases (e.g., oxygen, water vapor) have been reported for TLCP films.[44] Gas permeability studies by Paul and co-workers attributed low permeability of various gases initially to low solubility (vs. penetrant mobility), indicative of a highly crystalline or ordered substrate. Further examination of the effects of orientation and heat treatment indicate a moderate (45%) reduction in oxygen permeability with orientation and a lesser 10 to 20% reduction with heat treatment at SPP conditions (up to 24 h at 240°C under vacuum), driven in both cases by a reduction in penetrant mobility.[46] The results suggest that dense packing (i.e., interchain spacing) largely unaffected by macro-orientation or SPP governs the excellent TLCP barrier properties.

C. Uniaxial Sheets

TLCP fibers can be used as building blocks for larger uniaxial shapes due to their ease of self-adhesion and retention of orientation when heated in aligned bundles under pressure near (but below) the melting point, as indicated in several U.S. patents to DuPont.[47–49,56] For example, layups of uniaxial fiber warps can be converted to uniaxial sheets of 0.1 to 0.8-mm thickness on hot pressing with excellent retention of fiber tensile strength (~60%) and modulus (~75%) in the oriented sheet direction.[5,47–50] Moreover, the sheets are transparent, an excellent indicator of the monodomain character of the uniaxially oriented as-spun fibers retained in the hot-pressing process. Figure 2 is a photograph of a transparent TLCP sheet of 0.6-mm thickness taken from Reference 5, in which the fibrous character of the material is evident. Also shown are opaque marks made by localized melting (deorientation) with a heating iron.[5]

Sheets can be heat strengthened similar to that of a fiber to yield a novel 100% fiber resinless advanced composite, with tensile strengths to

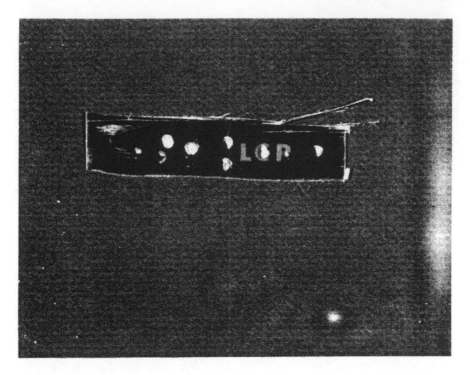

FIGURE 2
Transparent uniaxial TLCP Sheet. Oriented direction is horizontal. Opaque marks were made by localized melting (deorientation) with a heating iron. The word LCP shows through the sheet of 0.6-mm thickness. (From Luise, R. R., in *Integration of Fundamental Polymer Science and Technology-5*, Elsevier, New York, 1990, 207. With permission.)

1.9 GPa and moduli in the 50 to 60-GPa range.[5] The transparency is retained in the heat-strengthening step, another indication that the uniaxial morphology (i.e., interchain distance) is preserved on heat treatment, consistent with solid-phase polymerization. High-strength sheets can alternatively be laminated into multi-ply structures with conventional epoxy or TLCP film adhesives, or as previously mentioned, can be converted to biaxial film by transverse stretching on heating due to good lateral ductility.[5,47]

Table 4 compares the unidirectional tensile properties of the 100% TLCP heat-strengthened sheets with other uniaxial advanced composite materials (TLCP = poly-chloro-p-phenylene/biphenylene/terephthalate/isophthalate 90/10/90/10), taken from Reference 5. Of note is the exceptionally high strength and elongation (toughness) of the TLCP sheet combined with low weight vs. glass, Kevlar, and graphite/epoxy materials. Laminate properties (see Reference 5) are similar to those of Kevlar 49 composite materials indicative of low shear and compressive strength characteristics of these materials. The use of various cross-linking approaches applied to TLCPs such as oxidizing agents[20,23] and potential cross-linking comonomers such as sulfonate salts[51,52] or allyl-substituted aromatic diol intermediates,[53] could improve shear/compressional properties in fibers and sheet structures. Also, recent synthesis of liquid crystalline thermosets could provide routes to high-strength fibers and films or sheets with improved shear or transverse properties.[54]

TABLE 4.

Properties of Uniaxial Composites

	Tensile Strength (GPa)	Ultimate Elongation (%)	Tensile Modulus (GPa)	S.G.
100% TLCP, 0.6 mm	1.9	6.7–3.3	52–64	1.4
60% E-Glass/epoxy	1.2	2.9	43	2.1
60% S-Glass/epoxy	1.7	3.6	48	2.0
65% Kevlar 49/epoxy	1.4	1.8	76	1.4
60% H3501 Graphite/epoxy	2.1	1.5	150	1.5

From Luise, R. R., in *Integration of Fundamental Polymer Science and Technology-5*, Elsevier, New York, 1990, 207. With permission.

D. Pultruded Shapes

In a manner similar to uniaxial sheets, a U.S. patent to DuPont[56] shows that rovings of TLCP as-spun yarn can be pultruded by conventional methods into larger uniaxial shapes (e.g., rods, tapes) without a matrix binder. Significant improvements in tensile and shear mechanical properties are observed after heat-strengthening by methods similar to fibers.[5] Retention of fiber orientational properties is good. Mechanical properties comparable to other pultruded high-performance materials appear feasible, as shown in Table 5 for 1.32-mm rod taken from Reference 5 for the same TLCP as Table 4. Morphology is woodlike[5] but, as shown in Table 5, mechanical properties are superior to wood and to metals (aluminum, steel) on a weight basis. Low flex strength is attributed primarily to poor compressional properties, but may also indicate incomplete binding suggested by lack of transparency (vs. sheets). Tensile strength exceeded 1.0 GPa, with no break observed. The use of thermosets or other cross-linking approaches previously discussed for sheets could improve lateral properties. As TLCP prices decrease, long-range infrastructure applications are projected for these pultruded materials, e.g., in reinforced concrete and lightweight beams for bridges.[69] Sheets could have structural as well as optical applications.

E. Molded Articles

Injection-molded and extruded articles are also heat-strengthenable, as discussed in another U.S. patent to DuPont.[58,59] Improvements in tensile and flexural strengths (20 to 90%) are noted for standard plaques and

TABLE 5.

Properties of Pultruded Materials

	Tensile Strength (GPa)	Flexural Strength (GPa)	Tensile Modulus (GPa)	S.G.
100% TLCP, 1.3 mm	1.0[a]	0.3	41.9	1.4
70% E-Glass/epoxy	0.7	0.7	41.4	2.0
70% Kevlar 49/epoxy	1.4	0.6	82.8	1.3
Wood (maple)	0.1	0.04	12.4	0.9
Aluminum	0.3	0.3	70.0	2.7
Steel	0.7	0.7	210.0	8.0

[a] No tensile break (strength > 1.0 GPa).

From Luise, R. R., in *Integration of Fundamental Polymer Science and Technology-5*, Elsevier, New York, 1990, 207. With permission.

flex bars up to 3.2-mm thickness, and also for 1.2-mm extruded rods of copolyesters and polyazomethines. Heat treatment times shown are long, generally 24 to 36 h, with 12 to 16 h at the maximum temperature near the melting point (ca. 300°C). Toughness (or energy-to-break) and impact strength are most improved (100 to 200%) due to increased elongation-to-break, indicating the effectiveness of solid-phase polymerization even in these larger shapes. Strength improvements are generally lower than those of other more highly oriented uniaxial materials discussed above. Figure 3 is a plot of axial tensile strength response of various TLCP copolyester materials (fibers, uniaxial sheets, films, and moldings), showing the decline in strength/response with decreasing as-processed orientation.

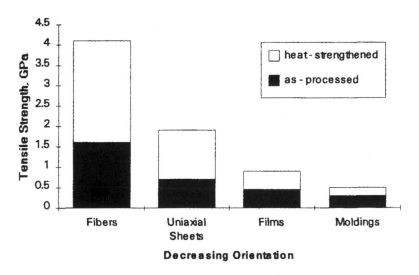

FIGURE 3
Axial tensile strength range for as-processed and heat-strengthened TLCP materials vs. as-processed orientation.

1. Mechanical Anisotropy

A number on injection-molding studies of TLCPs (mainly copolyesters) have noted the anisotropy of properties dependent on extensional flow-induced orientation.[18,59-64] The latter is found mainly in the surface of parts and is influenced principally by injection rate and melt temperature. As a consequence, mechanical properties (including dimensional stability) are generally higher in the flow (machine) direction than the transverse direction, which can pose design problems for certain parts. Fillers (e.g., chopped glass fibers) generally reduce the anisotropy and usually conveniently lower the cost, but recently melt-flow oscillation (i.e., disorienting) techniques have been developed which promise to improve mechanical property balance and eliminate weak weld lines. These processes include the SCORIM (Scortec), rheomolding (Solomat), and push-pull (Ferromatik Milacron) processes.[65]

F. Other Materials

A European patent application to Celanese discloses rigid TLCP foams with improved strength and thermal, flame, and solvent stability by heat strengthening.[66] Foams should be ideal SPP candidates since by-product removal should be facilitated. A U.S. patent, also to Celanese, discloses TLCP pulps composed of fibrils produced by masticating oriented as-spun fibers in a water slurry.[67] High orientation is preferred for high aspect ratio pulps, with fibrils in the 0.5- to 5-μm diameter range and L/D of 50 to 600 vs. 14- to 35-μm diameter for starting fibers. The pulps are heat strengthened in a nonoxidizing atmosphere which increases molecular weight and melting point, and also serves to bond fibrils during formation of a nonwoven webbed material. Similar effects are observed in nonwoven sheets prepared by random fiber laydown and heat-strengthening, as described in a U.S. Patent to DuPont.[70]

IV. SUMMARY

The purpose of this chapter has been to further clarify the industrial origins of TLCPs in the 1970s based on the published literature, and to review the unique post-heat-strengthening capability of these materials in various forms driven by solid-phase polymerization. While Carborundum and Eastman patents were issued earlier on fusible p-hydroxybenzoic acid-based copolyesters later characterized as liquid crystalline, a German patent application to DuPont published in 1975 appears to be the earliest literature recognition of thermotropicity in main-chain aromatic polyesters. The latter publication was also the first to disclose high-tenacity TLCP fibers by heat-strengthening.

Although heat-strengthening has been applied mainly to fibers and appears to be the basis for the Vectran high-tenacity TLCP fiber manufactured by Kuraray and Hoechst–Celanese, other TLCP product forms exist, often overlooked, in which heat strengthening has been effectively utilized. The latter includes film, uniaxial transparent sheets, and pultruded shapes prepared from fibers, injection-molded and extruded articles, and foams. While strength response is generally orientation dependent (fibers>films>moldings), other benefits such as enhanced thermal and solvent resistance due to increased melting point and molecular weight should have value in many non-fiber applications. Molded parts, in particular, should be ideal for large-scale heat-strengthening. In addition, larger uniaxial shapes such as pressed sheets and pultruded materials utilizing fibers as the basic building block should have important future reinforcement applications (e.g., infrastructures), particularly as TLCP costs decline and lateral bonding techniques are utilized to improve shear properties. Transparent sheets may also have optical display applications.

In terms of fundamentals, while a number of heat-strengthening studies discussed in this work tend to support a basic strengthening mechanism of postpolymerization (e.g., Yoon's tenacity model; unaffected lateral spacing indicated by film permeability and uniaxial sheet data), there is still a need for basic kinetic data for the solid-phase polymerization process. These include transesterification rates for copolyesters and the effects of thickness and orientation, often difficult to separate, on by-product diffusion and conversion rate. It is hoped that these will appear as TLCP commercial acceptance grows.

REFERENCES

1. Jackson, W. J., Jr., *Br. Polym. J.*, 154, December 1980.
2. Kirsch, M. A. and Williams, D. J., Understanding the thermoplastics business, *Chemtech*, 40, April 1994.
3. *Modern Plastics*, 54, January 1995.
4. Thomas, L. and Roth, D. D., Films from liquid crystals, *Chemtech*, 546, September 1990.
5. Luise, R. R., Liquid crystalline condensation polymers, in *Integration of Fundamental Polymer Science and Technology-5*, Lemstra, P. J. and Kleintjens, L. A., Eds., Elsevier, New York, 1990, 207.
6. Kwolek, S. L., Morgan, P. W., and Schaefgen, J. R., Liquid crystal polymers, in *Encyclopedia of Polymer Science and Engineering*, Vol. 9, 2nd ed., John Wiley & Sons, 1987, 1.
7. Calundann, G. W. and Jaffe, M., Anisotropic Polymers: their Synthesis and Properties, Proceedings of the Robert A. Welch Conference on Chemical Research, XXVI, Synthetic Polymers 1982.
8. Economy, J. and Volkson, W., Structural properties of aromatic polyesters, in *The Strength and Stiffness of Polymers*, Zachariades, A. E. and Porter, R. S., Eds., Marcel Dekker, New York, 1983, 293.
9. Chung, T.-S., *Polym. Eng. Sci.*, 36, no. 13, 901, July 1986.

10. Kleinschuster, J. J., Pletcher, T. C., Schaefgen, J. R., and Luise, R. R., German Offen. 2,520,820 (November 27, 1975); Belgian Patents 828,935 and 828,936 (November 12, 1975).
11. Pletcher, T. C., U.S. Patent 3,991,013 (November 9, 1976).
12. Kleinschuster, J. J., U.S. Patent 3,991,014 (November 9, 1976).
13. Schaefgen, J. R., U.S. Patent 4,118,372 (October 3, 1978).
14. Luise, R. R., U.S. Patent 4,183,895 (January 15, 1980).
15. Kuhfuss, H. F. and Jackson, W. J., Jr., U.S. Patent 3,778,410 (December 11, 1973).
16. Cottis, S. G., Economy, J., and Nowak, B. E., U.S. Patent 3,637,595 (January 25, 1972).
17. Jackson, W. J., Jr. and Kuhfuss, H. F., J. Polym. Sci., Polym. Chem., 14, 2043, 1976.
18. Jackson, W. J., Jr., Kuhfuss, H. F., and Gray, T. F., Jr., 30th Anniversary Technical Conference, February 4–7, 1975, Reinforced Plastics/Composites Institute, The Society of the Plastics Industry, Inc., Section 17-D, 1–4.
19. Calundann, G. W., U.S. Patent 4,067,852 (January 10, 1978).
20. Cottis, S. G., Economy, J., and Wohrer, L. C., U.S. Patent 3,975,487 (August 17, 1976).
21. Sawyer, L. C. and Jaffe, M., Structure-property relationships in liquid crystalline polymers, in High Performance Polymers, Baer, E. and Moet, H., Eds., Hanser, 1991, 68.
22. Calundann, G. W., U.S. Patent 4,067,852 (July 17, 1979).
23. Nakagawa, J., Kishino, Y., and Yamamoto, Y., U.S. Patent 5,045,257 (September 3, 1991).
24. Miramatsu, H. and Krigbaum, W. R., Macromolecules, 19, 2850, 1986.
25. Yang, D. K. and Krigbaum, W. R., J. Polym. Sci. Polym. Phys. Ed., 27, 1837, 1989.
26. Sarlin, J. and Tormala, P., J. Polym. Sci. Polym. Phys. Ed., 29, 395, 1991.
27. Volksen, W., Dawson, B., Economy, J., and Lyerla, J. R., Jr., Polym. Prepr., 20, no. 1, 86, 1979.
28. Yoon, H.-N., Strength of fibers from wholly aromatic polyesters, Colloid Polym. Sci., 268, 230, 1990.
29. Kwolek, S. L. and Luise, R. R., Macromolecules, 19, 1789, 1986.
30. Eskridge, C. H. and Luise, R. R., U.S. Patent 4,424,184 (January 3, 1984).
31. Gibbon, J. D., Lawler, T. E., Yoon, H.-N., and Charbonneau, L. F., U.S. Patent 4,574,066 (March 4, 1986).
32. Yoon, H.-N., EP 0,133,024 (February 13, 1985).
33. Muhlebach, A., Johnson, R. D., Lyerla, J., and Economy, J., Macromolecules, 21, 3117, 1988.
34. Blackwell, J., Biswas, A., Guitierrez, G. A., and Chivers, R. A., Trans. Faraday Disc. Chem. Soc., 79, 73, 1985.
35. Schmach, G. and Vogel, R., Kunst. Plast., 10, 82, 1992.
36. Johnson, D. J., Karacan, I., and Tomka, J. G., Polymer, 34, 1749, 1993.
37. Ide, Y., U.S. Patent 4,332,759 (June 1, 1982).
38. Ishii, T. and Sato, M., EP 0,346,926 (December 20, 1989).
39. Plastics Technol., 35, February 1991.
40. Toya, T., Hara, H., Iida, S., and Satoh, T., U.S. Patent 4,698,397 (October 6, 1987).
41. Toya, T., Hara, H., Iida, S., and Satoh, T., U.S. Patent 4,738,811 (April 19, 1988).
42. Toya, T., Hara, H., Iida, S., and Satoh, T., U.S. Patent 4,891,418 (January 2, 1990).
43. Watanabe, J., U.S. Patent 5,081,221 (January 14, 1992).
44. Thomas, L. D. and Roth, D. D., Chemtech, 546, September, 1990.
45. Chiou, J. S. and Paul, D. R., J. Polym. Sci. Polym. Phys. Ed., 25, 1699, 1987.
46. Weinkauf, D. H. and Paul, D. R., J. Polym. Sci. Polym. Phys. Ed., 30, 817, 1992.
47. Luise, R. R., U.S. Patent 4,786,348 (November 22, 1988).
48. Luise, R. R., U.S. Patent 4,939,026 (July 3, 1990).
49. Luise, R. R., EP 0,354,285 (February 14, 1990).
50. High Performance Textiles, 1, December 1990.
51. Shepherd, J. P., Farrow, G., and Wissbrun, K. F., U.S. Patent 5,227,456 (July 13, 1993).
52. Zhang, B. and Weiss, R. A., J. Polym. Sci., A, Polym. Chem., 30, 989, 1992.
53. Casagrande, F., Foa, M., Federici, C., and Chapoy, L. L., EP 0,383,177 (August 22, 1990).

54. Hoyt, A. E. and Benicewicz, B. C., *J. Polym. Sci. Polym. Chem. Ed.*, 28, 3403, 1990; 3417, 1990.
55. Blackwell, J., Cheng, H.-M., and Biswas, A., *Macromolecules*, 21, 39, 1988.
56. Luise, R. R., U.S. Patent 4,832,894 (May 23, 1989).
57. Kuraray Vectran Product Bulletin.
58. Luise, R. R., U.S. Patent 4,247,514 (January 27, 1981).
59. Huynh-Ba, G. and Cluff, E. F., Structural properties of rigid and semi-rigid liquid crystalline polyesters, in *Polymeric Liquid Crystals*, Blumstein, A., Ed., Plenum Press, 1985, 217.
60. Snokoc, E., Sarlin, J., and Tormala, P., *Mol. Cryst. Liq. Cryst.*, 153, 515, 1987.
61. Sweeney, J., Brew, B., Duchett, R. A., and Ward, I. M., *Polymer*, 33, 4901, 1992.
62. Boldizar, A., *Plast Rubber Process. Appl.*, 10, 73, 1988.
63. Bangert, H., *Kunst. Plast. (Munich)*, 79, 1327, 1989.
64. Menges, G., Schacht, T., Becker, H., and Ott, S., *Int. Polym. Process.*, 2, 281, 1987.
65. *Plast. Technol.*, 21, March 1994.
66. Ide, Y., EP 70,658, 1985.
67. Kastelic, J. R., Charbonneau, L. F., and Carter, T. P., Jr., U.S. Patent 4,395,307 (July 26, 1983).
68. Kaito, A., Kyotani, M., and Nakayama, K., *Macromolecules*, 23, 1035, 1990.
69. *Technol. News*, 29, January 1995.
70. Miller, P. E., U.S. Patent 4,362,777, 1982.

Chapter 3

EXTRUSION OF ORIENTED LCP FILM AND TUBING

Richard W. Lusignea

CONTENTS

I.   Introduction 41

II.   Background 42

III.   LCP Processing and Properties 46

IV.   Applications of LCP Film and Tubes 52
       A.   High-Barrier Containers 54
       B.   Electronic Packaging 55
       C.   Tubes and Lightweight Structures 57

V.   Conclusions and Summary 58

References 58

I. INTRODUCTION

Liquid crystal polymers (LCPs) were introduced commercially in the mid-1980s and have currently reached production levels of about 12 million pounds per year at a value of around $150 to $200 million per year. Almost all of their use is in injection molded parts such as connectors for electronic cables. Some production and use of LCP fibers is emerging, but is still small compared to other high-performance fibers. Major opportunities exist for LCP film and extruded parts to gain a significant share of markets, including over $500 million per year in high-barrier films, $450 million per year for electronic substrates, and $30 million per year for medical tubing, to name a few.

Two primary factors are blocking more widespread use of LCP materials in the marketplace: high material price and anisotropic properties. Both of these limiting factors can be overcome by processing which controls orientation and morphology, allowing relatively small amounts of LCP to be used cost effectively. New processing will spur greater use of LCPs which will increase demand and reduce prices.

This chapter will focus on processing and orientation of extruded film and tubing made from LCP resins, with emphasis on three market areas: high-barrier films, electronic packaging, and tubing for structures.

II. BACKGROUND

LCPs are so called because they can enter a state where the molecules are mutually aligned (crystal) yet the bulk LCP can flow (liquid). The fact that LCPs align in organized domains is a direct result of their high length-to-diameter ratio and rigid molecular structure. As the LCP flows, the local domains align in the flow directions, producing macroscopic fibrillar regions. Once the fibrillar regions are formed and oriented, their direction and structure persist, even when the LCP approaches the melt temperature. This can present problems if the orientation produces weak areas. However, it is a significant advantage if the orientation is in line with the direction of maximum stress. In this case, the LCP structure will be very efficient with regard to strength-to-weight and strength-to-cost ratios.

There are two major types of LCPs: those processed from the anisotropic melt state (thermotropic) and from anisotropic solutions (lyotropic). Figure 1 shows some typical LCP structures schematically. All LCPs are made up of linear rigid molecules that tend to align themselves in shear flow; the amount of molecular rigidity is determined by the chemical bonds. All commercial thermotropic LCPs are of the flexible-rigid copolymer type comprising a rigid (sometimes called mesogenic) monomer and other monomers that allow more chain conformations to control melt temperature, tensile properties, and other important features of the LCP. For example, Vectra® consists of hydroxybenzoic acid (HBA) as the mesogenic monomer and hydroxynaphthoic acid (HNA) as the flexible monomer.

At relatively low concentration solutions, lyotropic LCPs form an amorphous isotropic phase; then at a critical concentration the solution becomes anisotropic with local regions of alignment. When these locally aligned regions are organized globally, excellent properties can be obtained. Fiber spinning from anisotropic solutions is well known and is the process by which Kevlar® fibers are made. Indeed, fiber spinning and

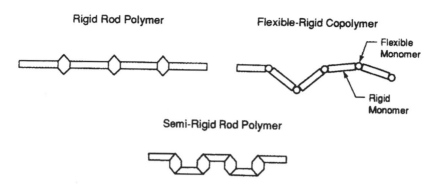

FIGURE 1
Types of LCP molecules.

other simple extrusion processes lead to very well-organized unidirectional LCP regions with a high degree of strength and stiffness in the
extrusion direction, but very little strength transverse to that direction.

The behavior of thermotropic LCPs is similar to lyotropic LCPs in
that both types form locally aligned regions in the liquid state. For thermotropic LCPs this is a molten state in which some of the LCP is aligned
and some is amorphous. The anisotropic melt state is similar to the anisotropic solution state, exhibiting shear thinning behavior wherein viscosity
and shear rate are inversely related, self-alignment of LCP molecules
occurs in shear flow, and the melt exhibits an extraordinarily long relaxation time of minutes to hours. Commercial thermotropic LCPs such as
Vectra® (Hoechst Celanese), Zenite® (DuPont), and Xydar® (Amoco Performance Products) all have an anisotropic melt state from which they
are processed. The LCPs must be processed properly in this melt state in
order to obtain desired properties. Tables 1 and 2 show the melt temperatures and typical properties of several commercial LCPs.

Obviously, LCPs are of useful interest in the solid state, where the
aligned domains have been locked in place, giving the LCP its distinctive
fibrillar morphology. Thermotropic LCPs reach the solid state simply by
cooling, which freezes in the microstructure. Processing of lyotropic LCPs
is more complex since the solvent must be removed. Typical solvents for
lyotropic LCPs are corrosive, relatively nonvolatile, strong acids; for example, sulfuric acid, methane sulfonic acid, and polyphosphoric acid.
Solvent removal requires a series of steps including coagulating the LCP
in a nonsolvent which results in a gel network of LCP fibrils. This is
followed by other steps to completely remove both the solvent and coag-
ulant, and to enhance the properties of the LCP. It should be clear that
lyotropic LCP processing is inherently more complex and expensive than
thermotropic LCP processing.

Since lyotropic LCPs require highly specialized processing, they are
generally available only as fibers made by companies that have invested
heavily in manufacturing plants to handle the LCP solutions. Thermotropic

TABLE 1.

Commercially Available LCPs Have Various Melt
Temperatures

LCP Trade Name	Melt Temp. (°C)
Xydar® SRT 900 (Amoco Performance Products)	420
Sumikasuper® E-6000 (Sumitomo Chemical)	330
Zenite® HX 6000 (DuPont)	320
Vectra® A-950	277
Vectra® B-950 (Hoechst Celanese)	—
Rodrun® 3000	195

TABLE 2.

Typical Properties of Biaxally Oriented Films Made from Thermotropic
LCPs by Superex Process

Property	Xydar® SRT-900	Vectra® A-950	Sumikasuper® E-6000
Tensile strength, ksi	19–28	24–34	29–40
(MPa)	(131–193)	(165–234)	(200–276)
Tensile modulus, Msi	1.1–1.2	1.2–1.8	1.3–2.9
(GPa)	(7.6–8.3)	(8.3–12.4)	(9.0–20.0)
Elongation at break, %	2–21	4–10	2–15

LCPs can be extruded, injection molded, and compression molded thermoplastically from pellets or powders using relatively inexpensive manufacturing methods. Accordingly, the remainder of this chapter will focus on processing and use of thermotropic LCPs, simply referred to henceforth as LCPs.

As shown in Table 1, commercial LCPs are made by at least five companies, and at least eight other companies are known to be making and selling relatively small quantities (less than 500 metric ton (t)/year each) of LCPs. Each type of LCP has a unique chemical formulation; and within each type there are various grades that have different melt temperatures, physical properties, and processing conditions which can be selected for application requirements to maximize performance and minimize cost. One supplier, Hoechst Celanese, has announced a blend of its LCP Vectra® with another thermoplastic, providing yet another way to meet the performance and price needs of market applications.

Some of the key properties of thermotropic LCPs are shown in Table 2, and Figure 2 compares the tensile strength and modulus of lyotropic and thermotropic LCPs with (non-LCP) polyester film. The lyotropic LCPs such as polybenzoxazole (PBO) have the highest mechanical properties because of their rigid rod molecular structure. Unfortunately, this also makes lyotropic LCPs difficult and expensive to process, as noted earlier. Thermotropic LCPs have true thermoplastic behavior with ability to be

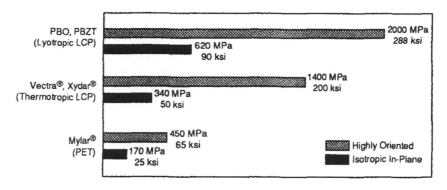

FIGURE 2(a)
Tensile strength of structural film.

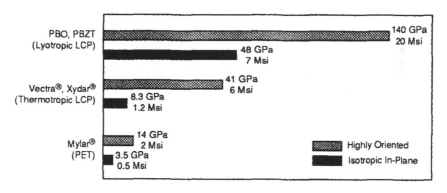

FIGURE 2(b)
Modulus of structural film.

used in a rich variety of processes, including recycling of the LCP parts by melt processing. A truly outstanding and useful property of LCPs is their low permeability to oxygen and water vapor, as shown in Figure 3. The high-barrier properties of LCPs often outweigh their increased cost over other materials, resulting in positive performance–cost tradeoffs. Because the oxygen barrier of LCP is six to eight times higher than ethylene vinyl alcohol (EVOH) at high humidity over 85% RH, but the price of LCP is four times higher, this results in a net materials cost savings of 33 to 50%. In other applications, LCP films and tubes have a combination of properties not found in other materials. For example, both high temperature capability over 150°C and high-barrier and chemical resistance make LCP the primary choice as film liners for reusable steel tanks that must be filled with hot corrosive materials.

Commercial LCP suppliers typically offer compounded LCP resins containing from 15 to 50% glass, mineral, or graphite fillers. Fillers tend to make the compounds more isotropic and less expensive, particularly

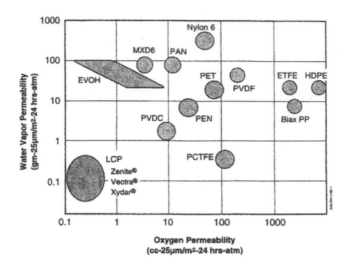

FIGURE 3
LCPs provide outstanding barrier properties.

for injection-molded parts. However, the fillers increase the density of LCP and degrade certain properties such as dielectric constant and loss factor. Certain LCP compounds can be extruded, but like the pure LCP resins, special processing is needed to achieve biaxial orientation. Compounds cannot be extruded into thin films (10 to 50 μm) since the size of the filler particles is very close to the thickness of the film. One very useful application for compounds is tubing which can be directly extruded with biaxial orientation, as described later.

LCPs can be used in polymeric blends and alloys to provide reinforcement to increase properties, or to reduce the viscosity of otherwise intractable thermoplastic resins. With regard to increasing properties, favorable performance-cost tradeoffs can be achieved when the improvement in properties is greater than the proportional increase in cost. This has been demonstrated for properties such as strength, stiffness, and gas barrier where as little as 10% LCP by volume has produced two to five times improvement in properties.[11-13] This results in material cost savings because the volume of polymeric material needed is reduced, which can provide additional cost savings through weight and size reductions. Also, thinner structures can reduce the effect of surcharges, frequently several times higher than the cost of the plastic itself, placed on the weight of plastic in a given structure.

III. LCP PROCESSING AND PROPERTIES

One of the greatest advantages of all polymeric materials is that they can be easily and inexpensively processed by extrusion, injection molding, thermoforming, and other familiar processes. Improved performance of random coil polymers can be achieved by orienting the molecules, which involves straightening and aligning the polymer chains. Products such as magnetic recording tape and poly(ethylene terephthalate) (PET) carbonated beverage bottles would not be possible without orientation. Material engineers decide to use special processing when the property improvements of oriented polymers offset the increased costs of processes such as tentering and stretch-blow molding.

When it comes to processing LCPs, there is really no decision between unoriented and oriented processing. LCPs will always be oriented by the effects of shear and elongational flows in processing. The major question in processing LCPs is, "How does one control the orientation direction where they will maximize properties and minimize the volume of LCP and cost of the finished part?" The importance of this question is further elevated by the relatively high cost of LCPs, which dictates efficient use to achieve favorable performance–cost tradeoffs.

Figure 4 shows two types of LCP film or sheet, the first with a skin–core morphology, the second with controlled orientation through the thickness. Using conventional processing, the skin–core morphology will generally result in a misoriented region of LCP that will reduce properties. New processing is needed to control the orientation, placing all the LCP material in the primary load directions.

Figure 5 shows how LCP film can be biaxially oriented with the counter-rotating die to place orientation at two principal directions of $\pm\theta$ to the machine (extrusion) direction. As shown in Figure 6, uniaxial orientation ($\theta = 0$) is produced by a stationary slot die, as opposed to the controlled biaxial orientation from the counterrotating die. The counterrotating die and other types of rotating extrusion dies take advantage of the LCP propensity to self-align in shear flow by controlling the direction of the flow. This places the LCP self-aligned domains in the most effective directions. Figures 7, 8, and 9 show the effect of orientation angle on strength, stiffness, and coefficient of thermal expansion (CTE) of biaxially oriented LCP film. The CTE for biaxially oriented LCP films with θ equal to 45° is much lower than other thermoplastics, typically –4 to +4 ppm/°C for LCP films. The low CTE is very useful in substrates for printed circuits with active silicon devices mounted on the substrates because the LCP film substrate will not create as much stress on the silicon chips as other polymer films when the temperature of the circuit changes. Processing with the counterrotating die shown earlier in Figure 5 gives the LCP film the necessary ±45° orientation in the plane of the film.

Other orientation methods have been tried with LCP films, but without much success. Both conventional blow molding and tentering result in LCP film with a much greater amount of orientation in the machine direction because of the natural tendency of LCP molecules to self-align and for this alignment to persist. In a typical blown film process, LCP melt is extruded through a stationary circular die which produces shear

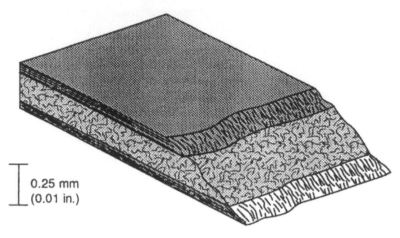

FIGURE 4(a)
Skin–core with conventional processing.

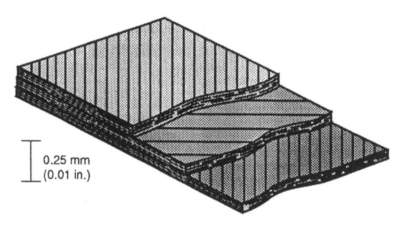

FIGURE 4(b)
High orientation throughout the thickness with new processing.

forces in the longitudinal direction as the melt flows axially between the walls of the die from inlet, to the die, to the exit gap. When the melt exits the gap, it will have a fibrillar orientation much like the diagram on the left side of Figure 6, and this orientation will tend to persist when the film is blown out radially by internal pressure, in a typical blown film process. The resulting film will have predominantly MD orientation, about four to five times higher than the TD orientation, even though relatively high blow out ratios (over 3:1) are used. Attempts to use higher blow out ratios have resulted in uneven weak spots in the bubble, blow outs, and inability to process the LCP film. Similar problems occur in a tentering process. Here the LCP melt is extruded through a stationary slot die, then stretched transversely to try to impart biaxial orientation. However, since almost

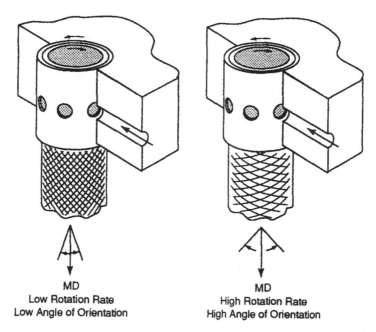

MD
Low Rotation Rate
Low Angle of Orientation

MD
High Rotation Rate
High Angle of Orientation

FIGURE 5
Counterrotating die.

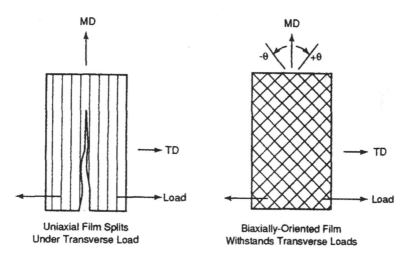

Uniaxial Film Splits
Under Transverse Load

Biaxially-Oriented Film
Withstands Transverse Loads

FIGURE 6
Molecular orientation of uniaxial and biaxial LCP film.

all of the LCP has been oriented in the machine direction because of one-way longitudinal shear in the die and this orientation is persistent, it will be very difficult to achieve significant transverse orientation. Transverse draw will tend to tear the film web before much cross-direction orientation can be achieved.

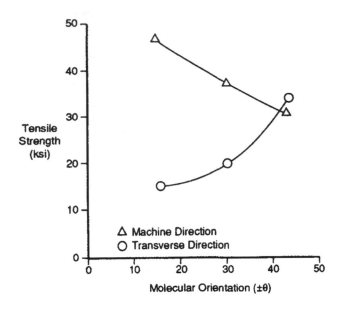

FIGURE 7
Producing LCP films with in-plane isotropy (tensile strength).

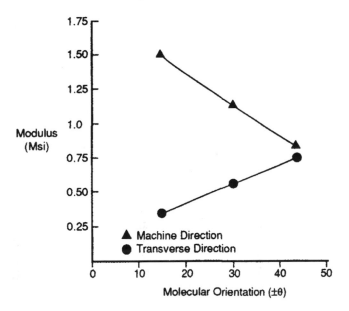

FIGURE 8
Producing LCP films with in-plane isotropy (modulus).

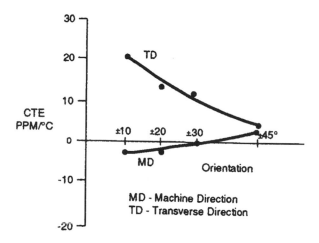

FIGURE 9
Producing LCP films with in-plane isotropy (CTE).

One of the key principles for making oriented film and tubing from LCPs is that the melt flow process must be controlled to achieve the desired orientation. Uncontrolled melt flow will probably result in the wrong type of orientation, or inefficient use of high-value LCP material.

Figure 10 shows another type of rotating die with three rotating cylindrical elements. This die allows for greater orientation through the thickness of LCP film, tube, and sheet; and for multiaxial orientation within the film or tube wall. This die has been used to make thin (25- to 100-μm) flat films from LCP for applications in multilayer printed circuit boards, and to make tubes with very high strength and modulus in both the longitudinal and hoop directions.

Rotating circular dies increase the fixed and operating costs of the blown film process, but should still result in economical processing of LCP film and tubing, especially with comparison to the alternatives, such as laminating several uniaxial LCP layers to get biaxial orientation. Work is being done on scaling up the rotating die systems; and although no hard data are available on prices, the rotating die system should be about 10 to 20% more expensive to build and about the same price to operate as a conventional blown film system.

The die shown in Figure 10 can also be used to coextrude oriented LCP layers with other thermoplastic layers. Recently, Superex has demonstrated that three-layer structures can be coextruded with such a die, resulting in PET-tie layer-LCP structures with total thickness of 25 to 50 μm and 30% LCP thickness. These films had excellent oxygen and water vapor barrier properties, as expected from the properties of LCP. Also, since only a thin layer of LCP is used, this multilayer film has very good performance–cost figures. In the future, it will be possible to coextrude

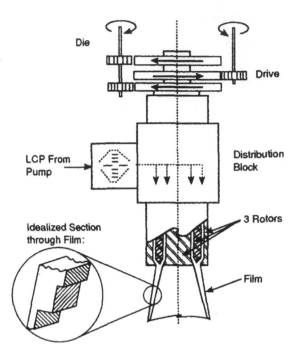

FIGURE 10
Trimodal die provides controlled orientation through film thickness.

LCP layers with other thermoplastics to make high-performance, cost-effective films.

The processing techniques discussed above for 100% LCP can also be used to advantage with filled LCP compounds and LCP-thermoplastic blends and alloys. The reason is that filled LCPs and LCP blends tend to self-align and show persistent orientation behavior, sometimes equal to or greater than the pure LCP. Tubing is a case in point where conventional processing was tried with various filled LCP compounds and resulted in tubing which split longitudinally when squeezed or flexed. The same LCP compounds were extruded into tubing using a rotating die and the resulting products were biaxially oriented and did not split, while retaining very good flexural stiffness and strength.

IV. APPLICATIONS OF LCP FILM AND TUBES

Applications for LCPs are being found for an increasingly wide variety of products, as shown in Table 3. As pointed out earlier, the price of LCP is currently high, about $12 to $22/lb, dictating LCP use in applications where performance outweighs price. Interestingly, this can mean some fairly high-volume commercial uses, such as food packaging and circuit boards for cellular phones. For example, in food packaging LCP can be used as a thin high-barrier layer to prevent oxygen from deteriorating the taste of precooked and packaged food. Since the barrier properties of the LCP are so good, a little bit goes a long way; this means that a layer of LCP less than 1/4 the thickness of polyvinylidene chloride (PVDC) or ethylene vinyl alcohol (EVOH) can do a better job, providing economic and performance benefits. In the area of printed circuit boards, LCP films can be made as inexpensively as thin (less than 50 µm) fiberglass epoxy, and have much better electrical properties, again providing economic and performance incentives for LCP use. With regard to tubing, LCP compounds can be extruded into biaxially oriented tubing in one step, resulting in tubes with tensile properties that rival fiber reinforced composite tubes which cost much more and are not as thermally or chemically resistant as LCP tubes. Once again, LCP tubes provide performance and cost benefits. These examples, summarized in Table 4, are under active product development at Superex because processing can be used to control orientation

TABLE 3.

LCP Product Applications

Fiber applications	Composite structures
	Ropes, cables
Film applications	High-barrier containers
	Electronic substrates
	Lightweight structures
Molded, net-shaped applications	Electrical connectors
	Precision plastic parts
Tubing	Surgical instruments
	Structures

TABLE 4.

Biaxially Oriented LCP Provides Performance and Cost Savings

Application	Current Competitive Materials	LCP Cost Savings for Equal Performance
Long shelf life, nonrefrigerated plastic container	EVOH barrier layer	20–40%
Thin substrate multilayer boards	Fiberglass epoxy	17–33%
	Aramid fiber epoxy	Over 50%
	Polyimide	40%
Tubing for endoscopic surgical instruments	Fiberglass-thermoset resin	About 50%

and make useful products which are competing in the marketplace. This is true even at today's high prices for LCP. As the price drops because of increased volume, supplier competition, and lower cost monomers, even more applications and use of LCP will occur.

A. High-Barrier Containers

The barrier properties of LCPs are excellent, as shown previously in Figure 3, making LCP tubes and films useful in the following products:

- Food containers which can be stored without refrigeration for microwave cooking
- Lids for food trays and cans, replacing metal lids which are heavy and have sharp edges, particularly for use with snacks and children's foods
- Multilayer bottles and tubes where a single LCP layer can replace several other barrier layers
- Liners for lightweight composite fuel tanks for compressed and liquefied natural gas
- Impermeable films for vacuum insulation panels
- Medical bags where LCP layers can replace PVDC layers, providing a bag which can be incinerated without concern about burning chlorinated plastic
- High-barrier multilayer bottles which can be made with thinner walls and less total material to reduce costs

LCPs can be used in thin layers or at low concentrations in blends to make cost-effective high-barrier films and containers. LCP barriers have advantages over aluminized plastics and metal foil because the LCP layers can be thermoformed, are microwaveable, recycling is easier, and metal detection systems can be used to inspect packaged foods.

Using biaxial film processing, LCPs can be made into thin liners for compressed and liquefied natural gas tanks, replacing thick and heavy high-density polyethylene (HDPE) liners or costly aluminum liners. An LCP liner only 0.051 mm thick provides ten times more barrier than an HDPE liner 9 to 13 mm thick, and does so at 1% of the weight and with a 30% savings in material costs compared to the HDPE liner. Superex is actively engaged in development of thin LCP liners, including blow molding and thermoforming to make the liner and filament winding to make the composite structure.

High-barrier film users are currently looking for alternatives to polyvinylidene chloride as a barrier layer because of problems cited in burning PVDC. LCP laminates and coextruded films are being evaluated because evidence suggests they can be burned as easily as PET, and their high-barrier properties mean that LCP layers will be four to five times thinner than PVDC for equivalent barrier performance.

Although ethylene vinyl alcohol (EVOH) is a very good oxygen barrier when dry, it has poor resistance and barrier to water vapor, and does not retain much strength at high temperatures used for food sterilization. For these reasons, manufacturers of packaged foods are looking into multilayer LCP film, sheets, and tubes for pouches, trays, and lids used in packaging. Superex has demonstrated the ability to coextrude a PET-tie layer-LCP film, as shown schematically in Figure 11.

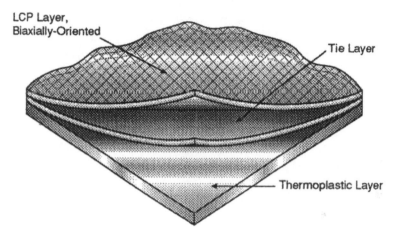

FIGURE 11
Schematic representation of Superex coextruded LCP-thermoplastic multilayer.

B. Electronic Packaging

Biaxially oriented LCP films have low in-plane CTE and low dielectric constant and loss factor, and they can be made thin (25 to 75 μm) with copper layers bonded to both sides. For these reasons, LCP films have the potential to replace fiberglass-epoxy laminates (FR-4) for a variety of new multilayer printed circuit board applications. At these low thicknesses, it is expensive to make and handle fiberglass fabrics and prepregs, whereas it is relatively easy to make and handle biaxially oriented LCP films. Also, the electrical properties of FR-4 materials are not as good as LCPs, and requirements for hand-held computers and wireless communication systems can be better met by LCPs. Since the dielectric constant of LCPs is low (2.7 to 3.0), and moisture has a negligible effect, LCPs can be used as thin layers less than 50 μm thick, placing circuit lines closer together than is possible with FR-4. This should result in up to 100 times greater circuit density with LCP than conventional fiberglass-epoxy circuit board materials, and at much lower cost than thin FR-4 or exotic materials such as fluoropolymers.

The circuit board shown in Figure 12 was made with ten signal layers on LCP circuit substrates, and has a controlled in-plane CTE of 8 ppm/°C.

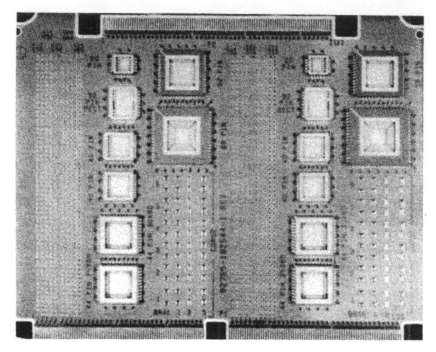

FIGURE 12
Ten-layer multilayer board made with biaxially oriented LCP film.

Superex is currently working on specific electronic packaging applications in the following systems:

- Thin multilayer boards for PC cards
- Transmit/receive modules for hand-held telephones and microwave circuits
- Flexible printed circuits with surface-mounted chips
- Plastic packages for silicon chips, providing hermetic quality

Alternative materials for these applications include polyimide, poly(ether ether ketone), and fiber-reinforced composites such as aramid paper epoxy and quartz polyimide. The prices of these materials is in the range of $50 to $220/lb; thus LCPs should compete very effectively at their current prices.

In addition to circuit substrates, LCPs and blends should find applications as dielectric layers in high temperature capacitors and as electrical insulation for power transmission and distribution cables and lightweight wire insulation. In a related application, LCP is being evaluated as a protective coating for optical fibers.

C. Tubes and Lightweight Structures

Uses are being found for biaxially and multiaxially oriented LCP tubing, as shown in Figure 13, in the following applications:

- Surgical instruments which must be nonconducting
- Lightweight structural tubing with low CTE
- Thin-walled, high-barrier tubing for fuel lines

FIGURE 13
Superex precision-extruded LCP tubing.

LCP tubes can compete effectively on a performance–cost basis with materials such as fiber-reinforced composites in tubes for surgical instruments. Since many new surgical procedures are done endoscopically with small incisions instead of open surgery, small instruments are needed by surgeons. Electrosurgical procedures require that these instruments be nonconducting to protect both the surgeon and the patient, which rules out the use of metal tubes. Plastic-coated metal tubes are expensive, and are subject to damage which could result in electrical failures. Fiber-reinforced composite tubes are relatively expensive; and, in one particular case, a biaxially oriented LCP tube could replace the composite at half the cost with equivalent mechanical properties.

V. CONCLUSIONS AND SUMMARY

Multiaxially oriented film and tubing can be processed from LCPs, compounds and blends using shear flow to orient the LCP molecular domains. Melt flow must be controlled to achieve desired directions of orientation and subsequent properties such as tensile strength and modulus, coefficient of thermal expansion, gas barrier properties, and electrical properties. When processed in thin-oriented layers or at low concentrations in LCP-thermoplastic blends, LCPs show performance and cost benefits over other types of film and tubing. Oriented film and tubing made from LCP can provide cost savings for the following reasons:

- LCPs have higher performance than their proportional increase in cost; thus they can reduce the total amount of polymeric material needed, which results in a direct cost reduction plus savings on surcharges that may be placed on the weight of material in a specific part.
- LCPs combine high temperature and chemical resistance with very good mechanical, electrical, and barrier properties; and can replace several different types of polymers with one LCP layer.
- Biaxially oriented LCP film and tubing are made by a single-step extrusion process which is inherently less costly than composite processing, so that LCPs can replace fiber-reinforced composites in certain price-sensitive applications.

LCP film and tubing can meet the performance and price requirements of quite a few applications at current LCP prices. As price per pound drops, large-volume markets in food and beverage packaging should open up for LCP multilayer film and tubing. Superex is working on near-term applications for the following:

- LCP tubing for endoscopic surgical instruments
- LCP substrates for circuit boards in small high-performance electronics
- LCP-thermoplastic multilayer film and sheet for food packaging
- LCP liners for lightweight compressed and liquefied natural gas tanks
- LCP insulation for electric wire and cable and optical fibers

REFERENCES

1. Ciferi, A., Krigbaum, W. R., and Meyer, R. B., *Polym. Liq. Cryst.*, Academic Press, New York, 1982.
2. Chapoy, L. L., *Recent Advances in Liquid Crystal Polymers*, Elsevier, New York, 1985.
3. Donald, A. M. and Windle, A. H., *Liquid Crystalline Polymers*, Cambridge University Press, U.K., 1992.

4. Osborn, K. R. and Jenkins, W. A., *Plastic Films Technology and Packaging Applications*, Lancaster, PA, Technomic Publishing, 1992.
5. Lusignea, R. W., Film processing and applications for rigid-rod polymers, *Mater. Res. Soc. Symp. Proc.*, 134, 265, 1988.
6. Ide, Y. and Ophir, Z., *Polym. Eng. Sci.*, 23, no. 5, 261, 1983.
7. Weinkauf, D. H. and Paul, D. R., in *Barrier Polymers and Structures*, ACS Symposium Series 423, 1990, 60.
8. Lusignea, R., Piche, J., and Mathisen, R., in *Polymeric Materials for Electronic Packaging and Interconnection*, ACS Symp. Ser., 407, 1989, 446.
9. Meier, M., The extrusion of liquid crystal polymers, SPE Symposium, ANTEC, New York, 1989, 200.
10. Farell, G. W. and Fellers, J. F., *J. Polym. Eng.*, 6, 263, 1986.
11. Blizard, K. G. et al., *Polym. Eng. Sci.*, 30, no. 22, 1442, 1990.
12. Blizard, K. G. et al., Processing of high performance dimensionally stable blends containing thermotropic LCP, ANTEC, Detroit, 1992, 2253.
13. LaMantia, F. P., *Thermotropic Liquid Crystal Polymer Blends*, Technomic Publishing, Lancaster, PA, 1993.
14. Williams, D. J., *Technical and Commercial Issues Governing the Development of Thermotropic Liquid Crystal Polyesters (TLCPs) in the 1990s*, PST Group, Harvard, MA, 1993.
15. Sun, T. et al., In situ formation of thermoplastic composites Ultem/Vectra®, *J. Composite Mater.*, 25, 788, July 1991.
16. Pawikowski, G. T. et al., Molecular composites and self reinforced liquid crystalline polymer blends, *Annu. Rev. Mater. Sci.*, 21, 159, 1991.
17. Malik, D. M. et al., Characteristics of liquid crystalline polymer polyester polycarbonate blends, *Polym. Eng. Sci*, 29, no. 4, 600, 1989.
18. Swkhadia, A. M. et al., Characterization and Processing of Blends of Polyethylene Terephthalate with Several Liquid Crystalline Polymers, ANTEC, New York, 1989, 1847.
19. Reisch, M. S., advanced thermoplastics producers regroup, *Chem. Eng. News*, 71, no. 35, 24, 1993.
20. Lusignea, R. W., Blizard, K., and Haghighat, R., Extrusion of High Temperature Thermoplastic Liquid Crystalline Polymer Microcomposites, Ultralloy '90 Symposium, Shotland Business Research, Houston, 1990, 151.
21. Anon., *Plast. News*, 46, 47, November 15, 1993.

Chapter 4

SCORIM: Principles, Capabilities, and Applications to Engineering and High Temperature Polymers

Eliot M. Grossman

CONTENTS

I. Principles of Operation 62

II. Equipment Description 62
    A.   The SCORIM Processing Head 62
    B.   The Hydraulic Drive Package (Power Pack) 64
    C.   The Controller 65

III. SCORIM Process 65

IV. Advantages of Using SCORIM and Test Results 66

V. Results and Discussion 68

VI. Applications 74

VII. High Temperature Materials 75
    A.   Testing and Applications 75

0-8493-7672-6/97/$0.00+$.50

VIII. Conclusions 76

References 77

I. PRINCIPLES OF OPERATION

SCORIM (Figure 1) is a new patented process employing novel equipment that, for the first time, creates dynamics to the molten plastics inside the mold cavity. This is accomplished by utilizing the SCORIM equipment that fits between the injection screw (plunger) and the mold.

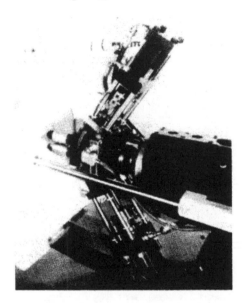

FIGURE 1
SCORIM process.

The concept of moving the molten plastic around inside the mold conjures up all kinds of interesting and useful solutions to existing problems in thin- as well as thick-sectioned parts to the molder, the plastics supplier, the product/part designer, the plastics and applications engineer, and the quality and reliability specialist.

These existing problems — internal stress relief, weld lines, sink marks, flow lines, warpage, voids, dimensional stability, weight, and density control and fiber orientation in reinforced plastics — are addressed and minimized, eliminated, or solved with the use of SCORIM technology.

II. EQUIPMENT DESCRIPTION

The SCORIM process utilizes the following three separate pieces of equipment.

A. The SCORIM Processing Head

This (Figure 2) is a specifically designed device that mounts at the end of the injection-molding machine and has three specific functions:

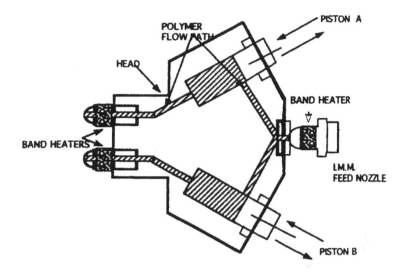

FIGURE 2
SCORIM processing head configuration.

1. The SCORIM head divides and melts from the barrel into two separate streams.
2. The SCORIM head heats and maintains the two streams at a selected melt temperature.
3. The SCORIM head acts as a pressure chamber and supports the platform for the hydraulic cylinders to force and move the melt streams into and through the mold cavity(s).

The mounting of the head to the barrel is accomplished with a number of different devices and mechanisms specifically selected to meet the customers needs and requirements, keeping in mind that alignment of the two nozzles to the mold is critical.

Options include, but are not limited to, direct mounting to the barrel cap, adapter, and support rings between existing barrel nozzle and the SCORIM head; adapter plates and a stub nozzle specifically designed to fit between the barrel and the head using minimum space.

The components that make up the SCORIM head assembly are (Figure 3):

1. The head body
2. The two hydraulic cylinders specifically sized to meet the customers needs that develop the pressures needed to apply to the melt

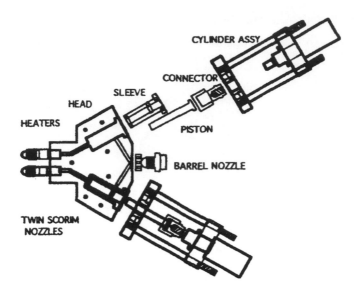

FIGURE 3
SCORIM head — exploded view.

3. The two couplings that join the piston rods and the hydraulic cylinders

4. The two piston rods that fit in the head and deliver the pressure directly to the melt

5. The two matching nozzles that mate the head to the mold sprue bushings and direct the melt into the mold

6. Cartridge and band heaters and thermocouples that develop and control melt temperatures in the head

B. The Hydraulic Drive Package (Power Pack)

This is a standardized hydraulic power delivery system that drives the cylinders on the head and with the sequencing valves controls the proper pressure, frequency, and duration of the SCORIM cycles as selected on the controller.

The power pack consists of the following components:

1. The drive motor(s)

2. The hydraulic pumps (either vane or gear)

3. The control and sequencing valves

4. The oil reservoir (60-gal capacity)

5. The electrical control and distribution box (220 V, 60A)

C. The Controller

This is a specifically designed electronic system that monitors, selects, controls, and limits the many functions of the SCORIM process. The basis and heart of the system is based on programmed logic control devices, and the functions that the controller monitors, selects, controls, and limits are head and nozzle temperature, cavity pressure (monitors only) mode selection (three modes, up to 12 repetitions), piston frequency in each mode, mode duration, piston pressure (forward and back), and total cycle. In addition, the controller provides a manual override as well as record-keeping capability (program storage).

The controller functions through a selectable menu presented on a user-friendly touch screen. Attached are the sequence of display screens which depict the simplicity of principle and selection.

III. SCORIM PROCESS

The SCORIM head is located between the injection screw and the mold, directly attached to the barrel (Figure 4). The head divides the main molten plastic stream from the screw into equal streams. Each stream passes through a piston chamber to its own nozzle, gate, and runner in the mold.

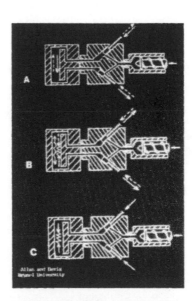

FIGURE 4
SCORIM head — Modes A through C.

The first step in the SCORIM process is to fill the mold either through one or both piston channels. (If the mold is filled through only one channel, one piston is positioned in the open position and the other is positioned in the closed or down position. The molten plastic is injected from the screw into the mold and on into the second channel where it meets the second piston. If the mold is filled through both channels, both pistons are positioned in the open position. The molten plastic is injected from the screw into the mold and the flow meets in the mold cavity.)

Once the mold is filled, the SCORIM cycle is initiated at the start of the machine hold cycle with the pistons moving in Mode A; that is, the pistons are gently moved back and forth out of phase for one or more cycles with screw pressure being maintained, depending on the part configuration and thickness.

Looking into the mold at the molten plastic during this action, one sees that a thin skin of solidified plastic has formed against the mold wall. The remaining molten polymer is forced by the actions of the SCORIM pistons to move backward and forward eliminating weld lines and their inherent weakness and flow lines. The movement of the pistons insured that the polymer in the gates remained molten, enabling further polymer to be introduced into the mold to compensate for and eliminate voids and sink marks as the polymer cools. In addition, the molten polymer is subjected to shear forces as it slides against the solidified layers through orienting any aligning and reinforcing fibers in the direction of movement.

The process then advances to Mode B where the pistons act together, in and out, for one or more cycles imparting compression/decompression forces, thus ensuring maximum and uniform packing while the polymer is still in a live state and the material is continuing to freeze in sequential layers toward the center of the part.

Final freeze of the part occurs while the process advances to Mode C where both pistons are in the down position for maximum compression.

It is of particular interest to note that when using unreinforced plastics, the SCORIM process has no adverse effect on cycle times; and when reinforced plastics are used, the cycle time could be extended by no more than 10%.

IV. ADVANTAGES OF USING SCORIM AND TEST RESULTS

What are the advantages of using the SCORIM technology and equipment; and where does the molder, part or product designer, materials supplier, and plastic part user gain?

This can be addressed by first describing what SCORIM technology and equipment can do (Figure 5).

Recalling that SCORIM technology and equipment is physically capable of moving the molten plastic around in the mold cavity; and remembering the basic definition of some of the classic molding defects, the ability to impart a dynamic movement to the molten plastic in the mold cavity should eliminate at best and certainly minimize their very existence at worst.

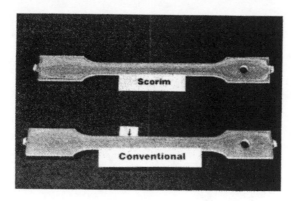

FIGURE 5
SCORIM technology moves the molten plastic around in the mold cavity, eliminating classic mold defects.

For example, a weld line or knit line by definition[1] is "The mark visible on a finished part made by the meeting of two flow fronts of plastic material during molding" and, structurally, is the weakest part of the molding (Figure 6).

When the molten plastic is manipulated and moved in the mold cavity, using SCORIM technology and equipment, the flow lines/marks are obliterated and a uniform distribution of molten plastic is created throughout the whole part.

Thus the flow lines/marks and inherent weakness are eliminated.

(It should be noted that because the plastic in contact with the mold surface, as the cavity fills, skins over rapidly, there may be the surface appearance of flow and/or weld lines, however, that are only "skin deep.")

A sink mark by definition[1] is "a shallow depression on the surface of an injection molded part created due to the collapsing of the surface due to local internal shrinkage after the gate seals."

When the molten plastic is manipulated and moved in the mold cavity, using SCORIM technology and equipment, particularly Modes B and C (referenced above), significant additional material under added packing pressure is introduced and held in the mold reducing the potential for sink marks to form since the gates are the last part to freeze off.

The sink marks then are, at a worst case, significantly minimized and totally eliminated.

A void by definition is "air or gas that has been trapped and cured into the molded part."

When the molten plastic is manipulated and moved in the mold cavity, using SCORIM technology and equipment, particularly Modes B and C (referenced above), significant additional material under added packing pressure is introduced and held in the mold thus squeezing out any entrapped voids.

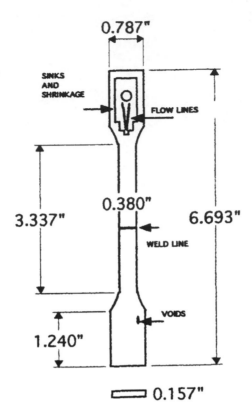

FIGURE 6
Molding defects corrected by SCORIM.

Voids then are significantly minimized if not totally eliminated.

Dimensional stability (Figure 7) by definition[1] is "the ability of a plastic part to retain the precise shape in which it was molded."

When the molten plastic is manipulated and moved in the mold cavity using SCORIM technology and equipment, internal molding stresses are relieved, thus minimizing the potential for shrinkage, warpage, or deformation. Because shrinkage, warping, and deformation in the molded part are minimized or eliminated, great dimensional stability is achieved.

Fiber alignment (Figure 8) by definition[1] is "the uniform orientation of reinforcing fibers within a polymer matrix in a given direction relative to a reference axis."

V. RESULTS AND DISCUSSION

When the molten plastic is manipulated and moved in the mold cavity using SCORIM technology and equipment, the shear forces in laminar flow induce a uniform alignment of the reinforcing fibers in the direction

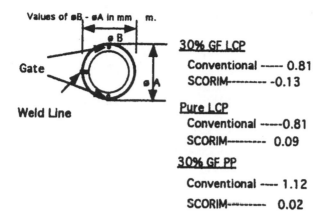

Values of øB - øA in mm   m.

Gate

Weld Line

30% GF LCP
Conventional ----- 0.81
SCORIM--------- -0.13

Pure LCP
Conventional -----0.81
SCORIM--------- 0.09

30% GF PP
Conventional ---- 1.12
SCORIM--------- 0.02

FIGURE 7
Dimensional stability [Typical Data (avg. of 10 moldings)].

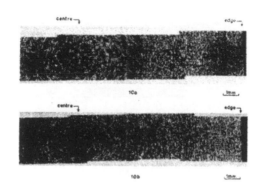

FIGURE 8
Fiber alignment.

of flow. Reinforcing fibers are then aligned or oriented in the direction of flow induced by the SCORIM process.

A variation of the two-head SCORIM process is the expanded four-head concept for use either with a single-screw molding machine or with a twin-screw machine.

The principles of operation are the same except that capability to alternate modes between any two pistons sequentially is possible so that a 0 to 90° layering effect can be achieved — a poor man's composite.

Previous studies have shown that reinforcing fillers generally have a very negative impact on weld line performance and strength. This is particularly true for short and long fiber reinforced materials. The butt weld strengths for the very highly reinforced polypropylene (PP) tended to be weaker than the strengths of the lower glass percentage materials when the samples were produced by conventional injection molding

(Figures 9 and 10). The samples molded with the SCORIM process showed dramatic property improvements over the conventional weld samples. The strength improvement is attributed to both weld healing or washout, and flow induced fiber (and molecular) orientation. The ultimate strength values were essentially equivalent for the three glass loading levels, although breaking strains decreased with increasing glass loading level. The SCORIM process pressures used did not seem to have a significant impact on the tensile properties. The results presented were molded at the 6.8-MPa pressure. While the strength increase with the SCORIM process was anticipated, the magnitude of the improvement was surprising as was the improvement in weld line appearance. It should be noted that previous research (unpublished) has shown that the optimum fiber loading for maximum mechanical properties is 42.3%.

The results for the mica-reinforced polypropylene (Figures 9 and 10) resin followed the same trend as the glass-reinforced material. The mica-loading levels evaluated were 0, 20, and 40%. The data presented in this

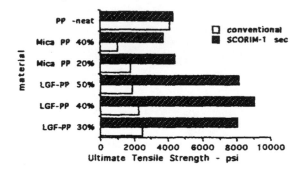

FIGURE 9
Ultimate tensile strength — PP various fills.

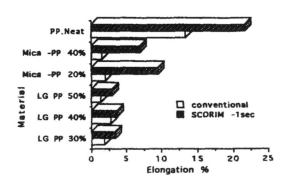

FIGURE 10
Ultimate tensile elongation — PP various fills.

overhead clearly show that the weld line tensile strengths decreased as filler percentage increased for the conventionally molded butt weld samples. The mica-filled samples did not yield and had relatively low-breaking strains. The weld line was very visible for these samples; and during testing, all samples broke cleanly at the weld interface.

The tensile properties for the mica-filled samples produced using the SCORIM process were significantly better than those produced conventionally. Both the tensile strength and the ultimate elongation improvements were significant. The SCORIM method seemed to effectively push material through the weld. The samples did not break in any one local area. Mixing of the mica seemed to be more uniform using SCORIM leading one to project that there might be advantages to be gained in uniform through part color mixing.

The next set of materials evaluated were the unfilled amorphous polymers (Figures 11 and 12). Two grades of high-impact polystyrene were run along with an ABS and a polycarbonate. For these unfilled materials, the tensile strength values were essentially the same for samples producing both the conventional and the SCORIM process. Yield strain values for the three materials were also essentially the same for each process; however, ultimate elongation values for the ABS and HIPS were significantly greater for the samples produced using the SCORIM process. Again, the elongation data for weld line samples must be regarded with caution due to the nonuniform strain distribution over the length of the part. The results do, however, indicate that the SCORIM samples are tougher in the test direction due to both weld line healing and orientation effects. The polycarbonate was somewhat unique in that the SCORIM process and conventional weld results were essentially the same. Both strength and elongation values remained essentially unchanged.

The next set of results presented in this chapter are for Vectra liquid crystal polymer (Figures 13, 14, and 15). The data for this material were generated by Hoechst–Celanese, Chatham, New Jersey. The material is of particular interest there because the butt weld line strengths for liquid

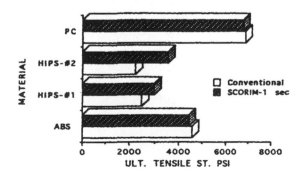

FIGURE 11
Ultimate tensile strength — amorphous materials.

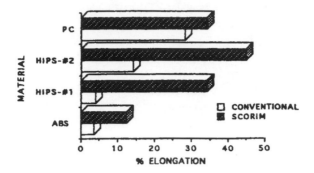

FIGURE 12
Ultimate tensile elongation — amorphous materials.

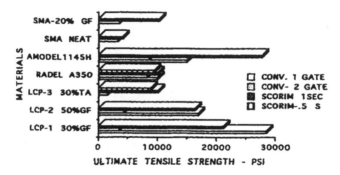

FIGURE 13
Ultimate tensile strength — engineered plastics.

crystal polymers can be quite low (low-strength retention factors) when samples are produced using the conventional injection-molding process. While these results are to be considered preliminary, the tensile properties of the double-end gated sample produced using the multilive feed process were significantly better than those of the conventional butt weld samples. The tensile strength values were, however, lower than those of conventional single-gate (nonweld) samples. Bevis[2] reports the SCORIM should enhance weld line strength to 85% of single gate-filled test bar as a minimum and in all cases that was exceeded. Unfortunately, the mold employed in this experiment had relatively heavy wall thickness (4.0 mm). Weld strength data for LCPs molded in thinner sections would be of more interest; however, the preliminary results presented here are encouraging, and indicate that additional weld line experimentation with these materials is warranted.

In addition, parallel data for Amodel, Radel, SMA, and Verton (long glass nylon, 50% filled) are presented showing significant improvement in tensile properties when SCORIM is used.

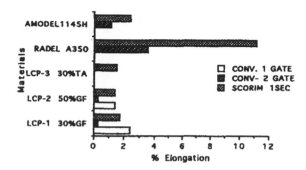

FIGURE 14
Percent elongation — various engineered plastics.

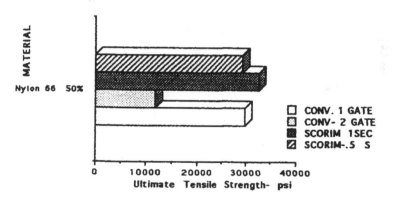

FIGURE 15
Verton nylon tensile strength.

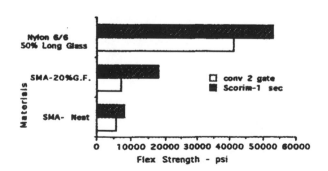

FIGURE 16
Flex strength.

First, indications show that comparable results are achieved when flex strength is evaluated using SCORIM (Figure 16).

VI. APPLICATIONS

Figures 17 and 18 illustrate some of the first real applications where SCORIM has contributed to significant product improvement.

Figure 17 is a VCR remote control cover with many holes, both round and rectangular. Weld lines were so prevalent that a paste on the insert was required to make the part aesthetically acceptable. In addition, the weld lines were a constant source of cracking and overall part failure. Use of SCORIM eliminated all weld lines, reduced costs by eliminating the

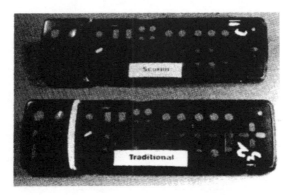

FIGURE 17
VCR remote control cover with many holes, both round and rectangular.

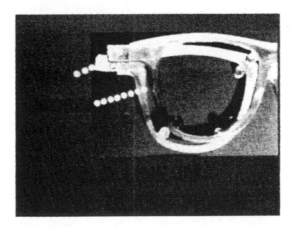

FIGURE 18
Molded eye glass frame.

paste on the insert, and in addition reduced overall molding cycles by 8 s (17%).

Figure 18 is a molded eye glass frame. Weld lines were presenting significant finish concerns and costs as well as a weak point in the frame where breakage and fracture could occur. Using SCORIM has given the customer a significantly stronger and more reliable part less prone to field failure and has reduced rejection rates and finishing costs.

VII. HIGH TEMPERATURE MATERIALS

A. Testing and Applications

As previously discussed and illustrated in Figures 13 through 16, liquid crystal polymers, Amodel, Radel, and long glass-filled nylon have been run with SCORIM and tested for tensile flexure and elongation properties and as illustrated show marked improvement in performance.

Liquid crystal polymers (LCP) are unique in their properties when manipulated with SCORIM. The movement of the molton polymer in oscillation between and through the gates induces long-chain crystal growth.

This is illustrated clearly in Figure 19 which shows 30% glass-filled LCP molded tensile bars. The left two samples show a two-gated tensile bar molded conventionally with a clean butt weld fracture at 3250 psi. The samples on the right were molded with SCORIM through two end gates. A thin skin of 0.010 to 0.015 in. capped the material that oscillated back and forth and as can be seen grew crystals of such length that they acted as fiber reinforcing. Tensile break occurred at 22,000 psi.

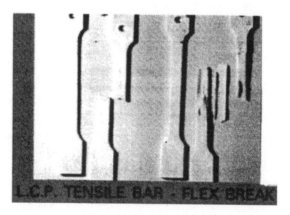

FIGURE 19
Glass-filled (30%) CP molded tensile bars.

Another example is shown in Figure 20. Rings molded with 30% glass-filled LCP and using four gates and then tested in tension.

The top ring illustrates the classic butt weld fracture while the bottom ring shows the 1-mg fiber-like crystal formations that contribute to tensile breaks at levels exceeding an order of magnitude above conventional molding.

FIGURE 20
Rings molded with 30% glass-filled LCP.

The future is with high-temperature polymers since work is now ongoing, showing promise using thermoset resins — SMC BMC, phenolic, and epoxies.

In addition, we at Scortec are looking at the potential for using LCPs as a reinforcing matrix with other polymers replacing glass and carbon.

VIII. CONCLUSIONS

The SCORIM apparatus showed an ability to significantly improve the weld line strengths of all the reinforced polymers used. The improvements in tensile properties are attributed to both fiber and molecular orientation, as well as weld line healing. In most cases, weld line tensile strength improvements obtained with unfilled polymers were small or negligible; however, improvements in ultimate elongation were generally realized with these materials. The SCORIM process has also been shown to improve the weld strengths for a liquid crystal polymer. The results presented indicate that this interesting new technology, which is in infancy, has a great deal of potential, particularly for reinforced polymers. Undoubtedly increasing usefulness will be found for SCORIM in the injection-molding industry as the potentials become known.

As previously mentioned and emphasized, weld line strength and failure is one of the serious concerns and limitations of the industry.

Elimination of the weld line with the resulting major mechanical strength and properties improvement as well as the aesthetic and cosmetic improvement is a quantum step forward in plastic part safety and reliability. It also opens the door to metal-to-plastics conversions that have not previously been considered viable.

REFERENCES

1. *The Engineered Material Handbook*, Vol. 2, Engineering Plastics, ASM International, Metals Park, Ohio, 1989.
2. Bevis, M., SPE Annual Technical Conference Proceedings, 1, 442, 1992.

Chapter 5

PROPERTIES OF LIQUID CRYSTALLINE THERMOSETS AND THEIR NANOCOMPOSITES

Elliot P. Douglas, David A. Langlois,
Mark E. Smith, Rex P. Hjelm, Jr.,
and Brian C. Benicewicz

CONTENTS

I.   Introduction 80

II. Synthesis 81
     A.   Amides.. 81
     B.   Esters 81

III. Thermal Behavior 82
     A.   Thermal Transitions 82
     B.   Phase Behavior.. 85
     C.   Thermal Stability.. 86

IV. Curing Behavior 87
     A.   Curing Kinetics of Bisacetylenes.................................. 87
     B.   Rheology of Propargyl Thermosets.............................. 90

V.   Nanocomposites 91

VI.   Conclusions 93

Acknowledgments 94

References 95

I. INTRODUCTION

Liquid crystal polymers (LCPs) have become an important class of materials over the last 20 years. Numerous studies have been devoted to understanding their synthesis, structure, morphology, and properties.[1,2] These advances have allowed liquid crystal polymers to become commercially important materials due to their high-strength, high-modulus, and good barrier properties. In particular, they are used extensively as fiber materials in advanced composites. By comparison, little attention has been paid to improvements in the properties of composite matrix materials. Typical matrix materials exhibit poor mechanical properties, usually being very brittle and having low strength. We have been investigating liquid crystal thermosets (LCTs) as improved matrix materials for use as high-performance structural materials. We have also investigated mixtures of LCTs with LCPs to create a new class of nanocomposites.

The LCT concept has been described previously[3-5] and is shown in Figure 1. The basic components of an LCT are a rigid core to which conventional cross-linking end groups are attached. Upon heating to elevated temperatures the LCT melts into a liquid crystalline phase. The end groups are then cured using conventional methods to give a solid matrix material with liquid crystalline order.

X = Conventional crosslinking group

Examples:

FIGURE 1
Schematic diagram of the liquid crystal thermoset concept. (Reprinted from Douglas, E. P., Langlois, D. A., and Benicewicz, B. C., *Chem. Mater.*, 6, 1925, 1994. With permission. Copyright 1994, American Chemical Society.)

The first description of liquid crystalline networks was given by de Gennes.[6] The first experimental reports on LCTs appeared in the early 1970s.[3,7] Interest was revived with the publication of several patents in the early 1980s.[8-12] Since then there have been several publications dealing with the synthesis and thermal behavior of LCTs.[13-16] However, there have been only a few reports on the physical properties of these materials.[17,18]

In this work we describe the behavior of LCTs that have been studied in our laboratory. We discuss the synthesis, thermal transitions, and curing behavior, as well as some preliminary results on physical properties. We also describe our use of LCTs to form nanocomposites.

II. SYNTHESIS

A. Amides

The general structure of the amide LCTs is shown in Figure 2. Materials based on maleimide, nadimide, and methyl-nadimide end groups have been prepared. Details of the synthesis have been published elsewhere,[4] but a general description is as follows: the end group is prepared by reacting the appropriate anhydride (for example, maleic anhydride in the case of the maleimide end group) with p-aminobenzoic acid. The resulting carboxylic acid is then converted to the acid chloride and reacted with an appropriate diamine under basic conditions to yield the amide LCT.

B. Esters

The general structure of ester LCTs is shown in Figure 3. A greater variety of esters has been prepared than of amides. Only a few representative examples are discussed here; more detailed discussions are available in our previous publications.[5,19] In all cases the esters are prepared by reacting two equivalents of an appropriate acid chloride with a diol, usually substituted hydroquinone. With suitable care these condensation reactions are quantitative. For the maleimide, nadimide, and methyl-nadimide end groups the acid chlorides are prepared in the same way as the amide LCTs. For acetylene LCTs the p-ethynyl benzoyl chloride was purchased from Ballard Advanced Materials, but could also be prepared according to the procedure of Melissaris and Litt.[20] For propargyl LCTs the end group is prepared by a modification of the procedure described by Dirlikov.[21] Methyl-4-hydroxybenzoate is reacted with a slight excess of propargyl chloride in the presence of potassium carbonate to yield methyl-4-propargylbenzoate. This product is then converted to the acid chloride in two steps. Alternatively, propargyl LCTs can be prepared by reacting propargyl chloride directly with a diol in the presence of

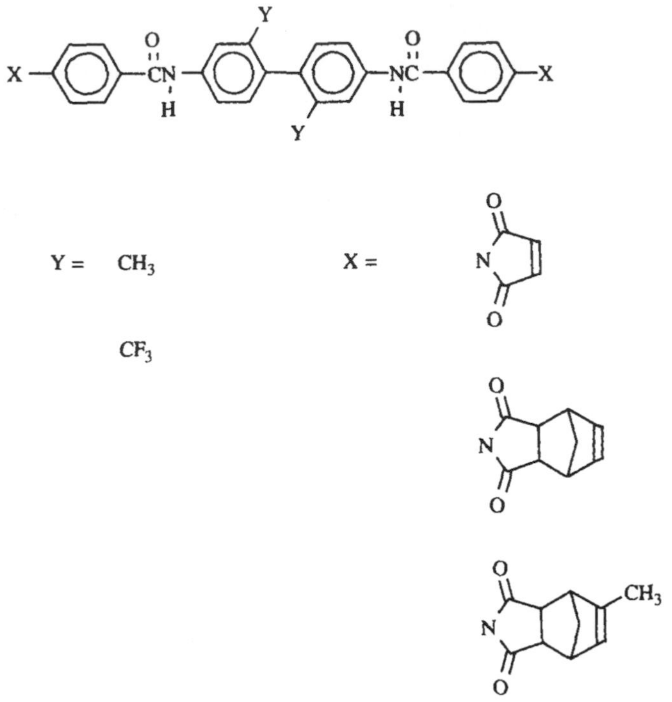

FIGURE 2
General structure of amide thermosets.

potassium carbonate. In this case the resulting thermoset has the propargyl group attached directly to the central mesogenic unit and does not contain an ester group. We have also prepared mixed propargyl thermosets in which one end is derived from the 4-propargylbenzoyl chloride and the other is derived from propargyl chloride, resulting in a thermoset with only one ester group.

III. THERMAL BEHAVIOR

A. Thermal Transitions

Thermal transitions in the LCTs have been studied by differential scanning calorimetry (DSC) and hot-stage polarized optical microscopy.[4,5,19] The DSC traces typically show zero, one or two endothermic transitions, followed at higher temperatures by an exotherm corresponding to reaction of the end groups. The number of endotherms and the types of transitions depend on several factors, including the bulkiness of any substituents and the nature of the end groups. Amides typically show

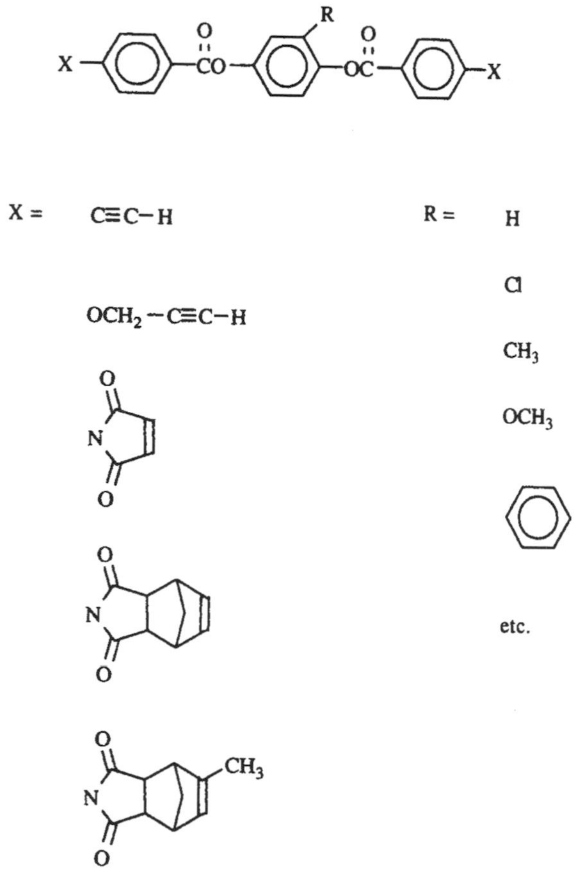

FIGURE 3
General structure of ester thermosets.

extremely high transitions or no transitions at all due to strong intermolecular interactions from hydrogen bonding. Esters show significantly
lower transitions than the amides. These results are summarized in Figures 4 and 5.

In Figure 4 can be seen the effect of the end group on the thermal
transitions. Upon changing the end group from maleimide to nadimide,
the stability of the nematic phase decreases as the end group becomes
more bulky. This is reflected in the appearance of a nematic to isotropic
transition for the nadimide thermoset that is not present for the maleimide
thermoset. Upon addition of the bulky methyl group to the nadimide
moiety, the stability of the nematic phase is decreased even further, as
evidenced by the lower temperatures for the transitions and the decreased
temperature range over which the nematic transition occurs. For the
propargyl thermoset the linkage between the propargyl group and the
phenyl ring of the core structure is considerably more flexible than the

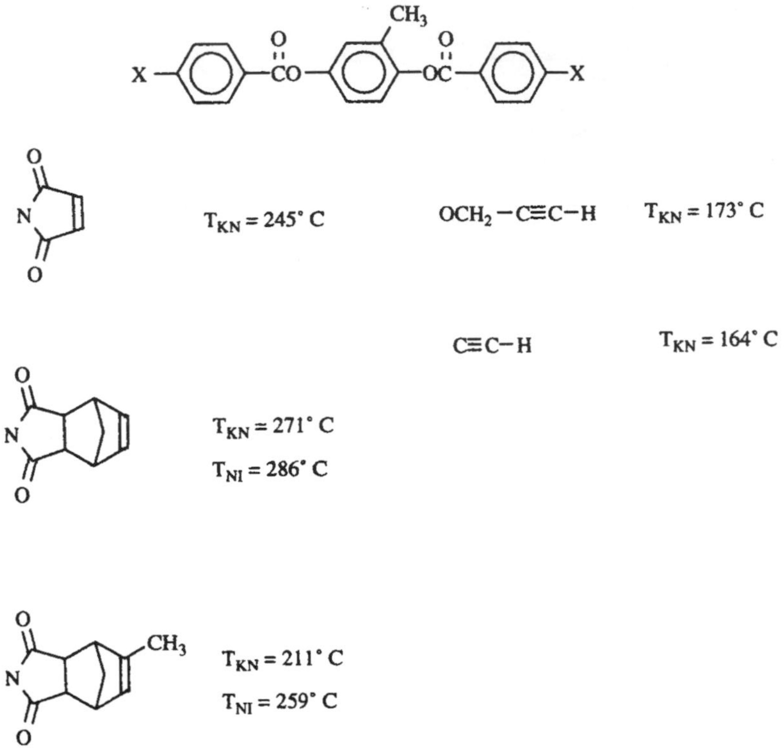
CH3
O
O
X
CO
OC
X
O
N
O
TKN = 245° C
OCH2 – C≡C – H
TKN = 173° C
C≡C – H
TKN = 164° C
O
N
O
TKN = 271° C
TNI = 286° C
O
CH3
N
O
TKN = 211° C
TNI = 259° C

FIGURE 4
The effect of changes in the cross-linking end group on the thermal transitions of liquid crystal thermosets.

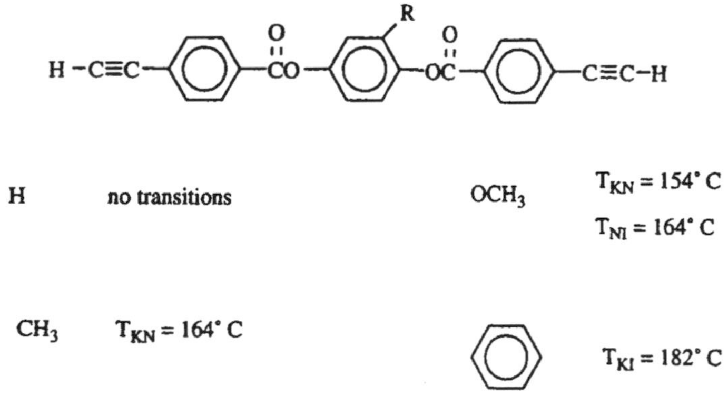
R
O
O
H – C≡C
CO
OC
C≡C – H
H
no transitions
OCH3
TKN = 154° C
TNI = 164° C
CH3
TKN = 164° C
TKI = 182° C

FIGURE 5
The effect of changes in the substituent on the thermal transitions of liquid crystal thermosets.

corresponding linkage for the imide groups, and the transition temperature is reduced accordingly. For the acetylene thermoset this linkage is also fairly flexible; however, the linearity of the acetylene group would suggest that the flexibility of the end group linkage should not play a role in affecting the transition temperature. We have previously found that electronic effects may be an important factor in how substituents on the central ring affect transition temperatures.[19] Similar electronic effects may also play a role with regard to the end groups.

Figure 5 shows the effect of substituents on the transition temperatures. Clearly, as the substituent gets larger the stability of the nematic phase decreases. We have found that there are three ways in which the substituent affects packing.[19] The most obvious effect is simply the size of the substituent. As the substituent becomes larger, lateral packing between molecules becomes less efficient and the overall order of the system decreases. We have also shown using molecular modeling that the size of the substituent affects the conformation of the molecule. Larger substituents tend to result in a conformation that is less linear along the length of the molecule, also reducing the packing efficiency and decreasing the order of the system. Finally, we have found that for very polar substituents the behavior is not explained by the above two effects. In these cases we attribute the behavior to electronic interactions that affect the packing of the molecules.

B. Phase Behavior

We have previously proposed a nonequilibrium phase diagram that describes the phase behavior of LCTs as a function of temperature and extent of cure.[5,19] This phase diagram is shown in Figure 6. The primary feature of this phase diagram is that it shows that the transition temperatures depend on the extent of reaction. The nematic transition is expected to increase with extent of cure due to an increase in the axial ratio of the molecules as the reaction proceeds. Larger axial ratios tend to promote the formation of ordered phases. The melting transition is expected to decrease with increasing extent of reaction since a mixture of different species is formed. The melting point then begins to increase again as a result of increased chain extension and cross-linking. Finally, at some point during the reaction a gel point is reached, beyond which the molecules are frozen and no further changes in the phase behavior occur. This is the point indicated by the vertical line in Figure 6. To the right of this line the material is a cross-linked solid.

We report here for the first time experimental confirmation of this nonequilibrium phase diagram. The results shown in Figure 7 were obtained by curing the bispropargyl LCT based on hydroquinone at 205°C

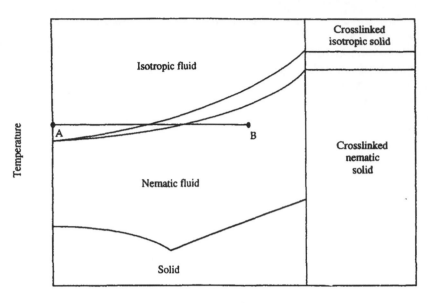

Extent of Reaction

FIGURE 6
Nonequilibrium phase diagram describing the behavior of liquid crystal thermosets. (Reprinted from Douglas, E. P., Langlois, D. A., and Benicewicz, B. C., *Chem. Mater.*, 6, 1925, 1994. With permission. Copyright 1994, American Chemical Society.)

on the hot stage of an optical microscope and examining the phase behavior as a function of cure time. The experimental results confirm the qualitative predictions of our model.

As a result of the increase in clearing temperature with extent of reaction, the possibility exists for a thermoset to initially melt into an isotropic phase, and for a nematic phase to subsequently form as a function of the extent of reaction. In this case the thermoset would follow the reaction path from A to B shown in Figure 6. We have, in fact, observed this behavior in several of our systems.[5,19]

C. Thermal Stability

Preliminary measurements have been performed on the thermal stability of the bispropargyl thermosets using thermogravimetric analysis (TGA). The cured materials exhibit excellent thermal stability. For example, when heated at 10°C/min in air the bispropargyl thermoset based on methyl hydroquinone shows initial weight loss at 290°C. Isothermal studies show no weight loss after 5 h at 260°C.

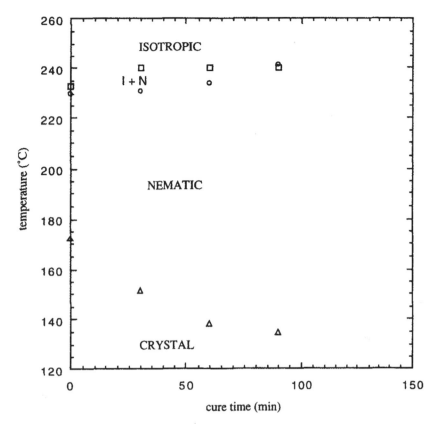

FIGURE 7
Experimental phase diagram for the thermoset of Figure 3 in which X is propargyl and R is hydrogen. Transitions were determined by heating and cooling on the hot stage of a polarized optical microscope at 10°C/min.

IV. CURING BEHAVIOR

A. Curing Kinetics of Bisacetylenes

We have performed extensive studies on the curing kinetics of the bisacetylene thermosets.[19] Our initial motivation was to help resolve some of the controversy over whether reactions that take place in a liquid crystalline phase occur faster than reactions in an isotropic phase.[22,23] To that end we have synthesized isomers of the thermosets based on either substituted hydroquinone or substituted resorcinol. Examples of these structures are shown in Figure 8. Clearly the resorcinol-based thermoset, with its bent conformation, is isotropic.

Curing kinetics were studied using Raman spectroscopy. Spectra were normalized to the absorbance of the carbonyl band at 1735 cm⁻¹, and then the absorbance of the acetylene band at 2108⁻¹ was followed as a function of cure time. Typical results are shown in Figures 9 and 10, which illustrate the curing of the thermosets of Figure 8 at 190°C. As can be seen in

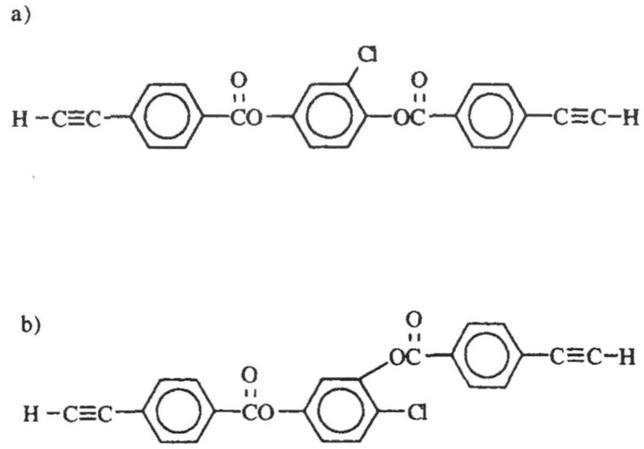

a)

b)

FIGURE 8
Examples of the thermosets used for curing studies. The structure in a) is nematic, and the structure in b) is isotropic.

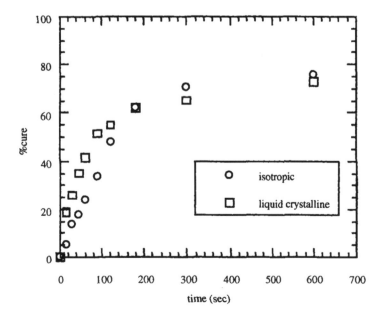

FIGURE 9
Curing curves for the thermosets shown in Figure 8. (Reprinted from Douglas, E. P., Langlois, D. A., and Benicewicz, B. C., *Chem. Mater.*, 6, 1925, 1994. With permission. Copyright 1994, American Chemical Society.)

Figure 9, curing of the bisacetylene thermosets is extremely fast. For the data shown, the materials are over 60% cured after only 3 min. Figure 10 shows the initial portion of the curves in Figure 9 in order to illustrate the rate enhancement in the liquid phase compared to the isotropic phase. We have previously used linear fits to describe the data of Figure 10.[19] Using this type of analysis we have concluded that there is an enhancement in the overall rate of cure for the liquid crystalline thermoset over the isotropic thermoset. Alternatively, the data can be fit as is shown in Figure 10. This analysis leads to the conclusion that there is a faster induction period for the liquid crystalline thermoset compared to the isotropic thermoset. Due to the very fast rate of reaction of the acetylene end groups, we are not able to obtain the data at shorter times that are necessary to distinguish between these two possibilities. In either case, however, curing clearly occurs faster in the liquid crystalline phase. This rate enhancement occurs due to the order in the liquid crystalline phase. In order for two reaction sites to come into proximity with each other, diffusion in a nematic phase only requires translation of one rod parallel to its neighboring rods. Clearly this type of diffusion will be considerably faster than diffusion of a rod through a random orientation of neighboring rods, as is the case for an isotropic phase.

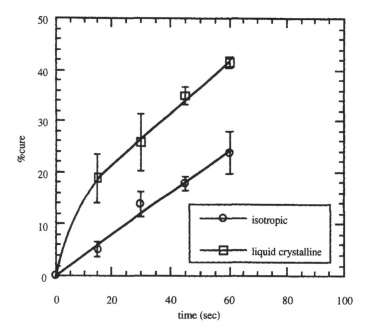

FIGURE 10
Initial region of the curing curves shown in Figure 9. (Reprinted from Douglas, E. P., Langlois, D. A., and Benicewicz, B. C., *Chem. Mater.*, 6, 1925, 1994. With permission. Copyright 1994, American Chemical Society.)

B. Rheology of Propargyl Thermosets

Propargyl thermosets require long cure times. A typical cure cycle for thermal cure is precuring for 16 h at 185°C followed by cure for 4 h at 208°C.[21] These long cure times make rheological measurements an appropriate technique for characterizing their cure behavior.

Figure 11 shows a typical cure curve for bispropargyl thermosets. The temperature increase to 208°C occurs at the point at which the G' and G" curves begin to increase rapidly. Clearly the precuring step does not result in the formation of any significant polymeric structure. It has been reported that curing of propargyl resins occurs by formation of a chromene ring followed by polymerization of the chromene ring.[21,24] The precuring is required to form the chromene ring. Curing at higher temperatures results in polymerization through the chromene ring to form a network structure. With this cure schedule, the gel point as measured by the crossover of G' and G" takes approximately 16.5 h, or 30 min beyond the temperature increase.

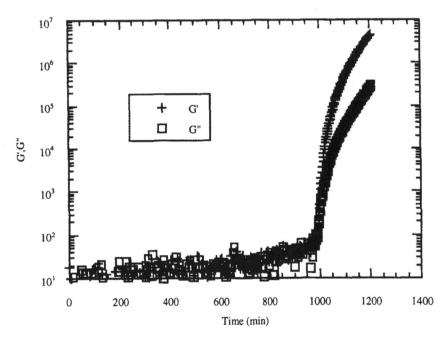

FIGURE 11
Storage and loss moduli during cure for the thermoset of Figure 3 in which X is propargyl and R is hydrogen. The temperature was held at 185°C for the first 960 min, and then increased to 208°C for the remainder of the cure. The frequency of the measurement was 1 Hz.

Use of a catalyst can substantially reduce this cure time. For example, when bis(triphenylphosphino)palladium dichloride is used as a catalyst, G' begins to increase after approximately 8.5 h, even without increasing the temperature to 208°C. An additional increase in G' is seen after 16 h, when the temperature is increased to 208°C. This behavior is illustrated in Figure 12.

We have also made preliminary measurements on the viscoelastic properties of the bispropargyl thermosets. An example for a liquid crystalline thermoset is given in Figure 13. The α relaxation (glass transition) is 257°C. It is noteworthy that the temperature of this transition is higher than the final cure temperature. There is also a very strong, symmetrical β relaxation at low temperatures. The symmetry of this relaxation is attributed to efficient packing in the liquid crystalline phase.

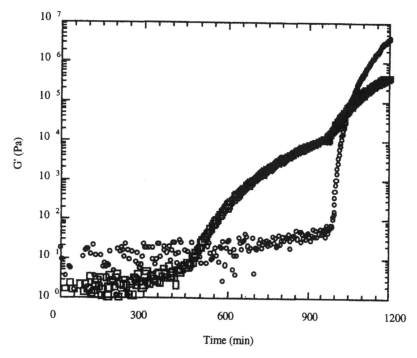

FIGURE 12

Comparison of catalyzed and uncatalyzed cure for bispropargyl thermosets. The thermoset used and the measurement conditions are the same as for Figure 11.

V. NANOCOMPOSITES

In addition to studying the properties of the neat LCTs, we are investigating their use in the formation of unique molecular or nanocomposites.[25,26] The basis of this type of composite lies in the concept of using a single liquid crystal polymer (LCP) molecule as the reinforcing element in a polymer matrix.[27] Since composite theory predicts that modulus and strength increase with increasing aspect ratio of the reinforcing fiber, the extremely high aspect ratio of a rigid rod polymer molecule is expected to lead to significant increases in composite properties.[28]

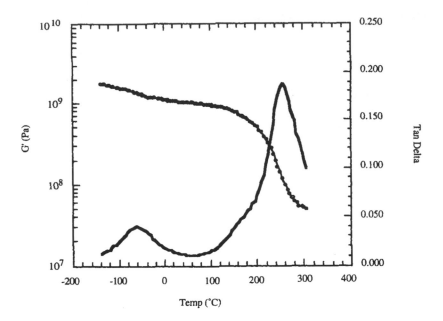

FIGURE 13

Storage modulus and tan δ for the cured thermoset of Figure 3 in which X is propargyl and R is hydrogen. The frequency of the measurement was 1 Hz.

Clearly, the composite approach outlined above requires a homogeneous mixture of the LCP in a matrix. Previous approaches have used coil (isotropic) polymers as the matrix component.[29] However, the entropy of mixing between a rod and a coil favors phase separation of the two components.[30] Consequently we have utilized mixtures of LCTs and LCPs in an attempt to overcome this problem.

The mixtures we have studied consist of the LCP poly(p-phenylene nitroterephthalamide) and the LCT of Figure 2 in which Y is methyl and X is maleimide. These mixtures are prepared from solutions of N-methylpyrrolidinone (NMP). Films are prepared by evaporation of the solvent on a glass plate, and are typically 20 to 50 μm thick. Fibers are prepared by dry-jet wet-spinning on a laboratory-scale fiber spinner. Initial solutions for spinning are approximately 10 wt% solids content. The resulting fibers are typically 40 to 70 μm in diameter.

Our initial studies utilized small angle neutron scattering (SANS) to study the details of the phase structure of cast films.[25] Most notably, we have found that domains form which are approximately 80 Å in size. The shapes of the domains vary depending on the composition of the mixture, being spheroid for a 100/0 LCP/LCT mixture, rodlike for an 80/20 mixture, and a randomly interpenetrating network structure for a 60/40 mixture. In the case of the 60/40 mixture, the scattering results suggest that there is only weak segregation of components between the two phases.

Given that these composites are prepared from solution, it is important to understand the structure development during the drying process. This type of knowledge allows control of the processing to form desired structures. Abe and Flory[31] have predicted that mixtures of rods of differing length in solution will exhibit phase separation into long rod-rich and short rod-rich domains based on the entropy of mixing. This result suggests that if the gel point during the drying process occurs in a biphasic region of the phase diagram, the desired homogeneous structure will not be obtained. In this case the final structure of the dried material is in a nonequilibrium state and reflects the structure that was present at the gel point.

In order to understand the structural evolution during the drying process, we are investigating the ternary phase diagram of LCP/LCT/NMP mixtures. The phase diagram based on polarized optical microscope observations is shown in Figure 14. A biphasic gap is seen which is qualitatively similar to the theoretical predictions, although the position of the experimentally observed gap is shifted significantly toward the solvent corner. This may be due to several omissions in the theoretical calculations, most noticeably the absence of chain flexibility and of enthalpic interactions.

We have performed some preliminary experiments to probe the nanoscale structure of these solutions.[26] Our initial results indicate that the nanoscale structure is significantly different from and more complex than the macroscale structure as described above. We are continuing these experiments with the aim of understanding how the structure of the dried films evolves from the initial solutions.

VI. CONCLUSIONS

Liquid crystal thermosets offer the possibility of advances in polymer matrix materials comparable to the advances seen over the last 20 years in polymer fibers. The LCTs we have described are easily synthesized in quantitative yield using simple condensation reactions. The curing rate can be readily controlled by changes in the cross-linking end group and any substituents. For example, curing times can range from a few minutes for the bisacetylenes to several hours for the bispropargyls. The temperature range over which a liquid crystalline phase occurs can also be controlled by changes in both the end group and any substituents.

There are several advantages of LCTs over conventional thermosets. We have shown here that the curing rate for liquid crystalline thermosets is substantially enhanced over the curing rate for isotropic thermosets. Even if all other properties are equal, this enhanced curing rate may prove advantageous, resulting in shorter cycle times during processing. We have also shown the possibility of using LCTs in novel nanocomposites. Used

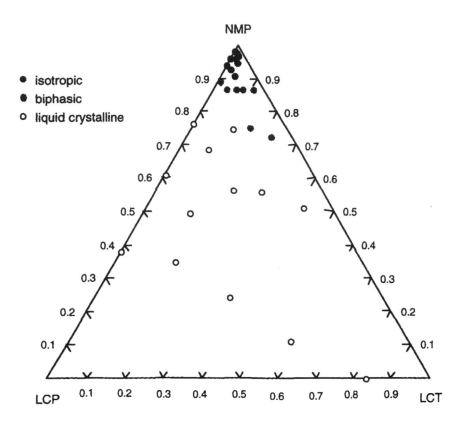

FIGURE 14

Ternary phase diagram of LCP/LCT/NMP mixtures based on polarized optical microscopy at room temperature.

either alone or in combination with other materials, liquid crystalline thermosets provide the opportunity for improved high-performance materials for applications in which high strength and light weight are the critical factors.

ACKNOWLEDGMENTS

The neutron scattering measurements were performed on three instruments: LQD at the Manuel Lujan Neutron Scattering Center, Los Alamos National Laboratory, and SAD at IPNS, Argonne National Laboratory, which are both supported by the U.S. Department of Energy; and NG7SANS at the National Institute of Standards and Technology, which is supported by the U.S. Department of Commerce. We thank P. Thiyagarajan (IPNS) and C. Glinka (NIST) for their expert help in conducting the neutron scattering measurements. This work was supported by the U.S. Department of Energy under contract W-7405-ENG-36 to the University of California.

REFERENCES

1. Ciferri, A., Krigbaum, W. R., and Meyer, R. B., *Polymer Liquid Crystals*, Academic Press, New York, 1982.
2. Ciferri, A., *Liquid Crystallinity in Polymers: Principles and Fundamental Properties*, VCH Publishers, New York, 1991.
3. Clough, S. B., Blumstein, A., and Hsu, E. C., Structure and thermal expansion of some polymers with mesomorphic ordering, *Macromolecules*, 9, 123, 1976.
4. Hoyt, A. E. and Benicewicz, B. C., Rigid rod molecules as liquid crystal thermosets. I. Rigid rod amides, *J. Polym. Sci. Polym. Chem. Ed.*, 28, 3403, 1990.
5. Hoyt, A. E. and Benicewicz, B. C., Rigid rod molecules as liquid crystal thermosets. II. Rigid rod esters. *J. Polym. Sci. Polym. Chem. Ed.*, 28, 3417, 1990.
6. de Gennes, P. G., Possibilites offertes par la reticulation de polymeres en presence d'un cristal liquide, *Phys. Lett.*, 28A, 725, 1969.
7. Liebert, L. and Strzelecki, L., Copolymerisation des monomères mésomorphes dans la phase nématique en présence d'un champ magnétique, *C. R. Acad. Sci. Paris*, 276, 647, 1973.
8. Conciatori, A. B., Choe, E. W., and Farrow, G., U.S. Patent 4,440,945, 1984.
9. Conciatori, A. B., Choe, E. W., and Farrow, G., U.S. Patent 4,452,993,1984.
10. Conciatori, A. B., Choe, E. W., and Farrow, G., U.S. Patent 4,514,553, 1985.
11. Calundann, G. W., Rasoul, H. A. A., Hall, H. K., U.S. Patent 4,654,412, 1987.
12. Stackman, R. W., U.S. Patent 4,683,327, 1987.
13. Litt, M. H., Whang, W., Yen, K., and Qian, X., Crosslinked liquid crystal polymers from liquid crystal monomers: synthesis and mechanical properties, *J. Polym. Sci. Polym. Chem. Ed.*, 31, 183, 1993.
14. Melissaris, A. P. and Litt, M. H., New high-Tg, heat-resistant, cross-linked polymers. I. Synthesis and characterization of di-*p*-ethynyl-substituted benzyl phenyl ether monomers, *Macromolecules*, 27, 883, 1994.
15. Melissaris, A. P. and Litt, M. H., New high-Tg, heat-resistant, cross-linked polymers. II. Synthesis and characterization of polymers from di-*p*-ethynyl-substituted benzyl phenyl ether monomers, *Macromolecules*, 27, 888, 1994.
16. Melissaris, A. P. and Litt, M. H., High modulus and high Tg thermally stable polymers from di-*p*-ethynylbenzoyl ester monomers: synthesis, solid state polymerization, processing, and thermal properties, *Macromolecules*, 27, 2675, 1994.
17. Barclay, G. G., McNamee, S. G., Ober, C. K., Papathomas, K. I., and Wang, D. W., The mechanical and magnetic alignment of liquid crystalline epoxy thermosets, *J. Polym. Sci. Polym. Chem. Ed.*, 30, 1845, 1992.
18. Barclay, G. G., Ober, C. K., Papathomas, K. I., and Wang, D. W., Rigid-rod thermosets based on 1,3,5-triazine-linked aromatic ester segments, *Macromolecules*, 25, 2947, 1992.
19. Douglas, E. P., Langlois, D. A., and Benicewicz, B. C., Synthesis, phase behavior, and curing studies of bisacetylene rigid-rod thermosets, *Chem. Mater.*, 6, 1925, 1994.
20. Melissaris, A. P. and Litt, M. H., Economical and convenient synthesis of *p*-ethynyl-benzoic acid and *p*-ethynylbenzoyl chloride, *J. Org. Chem.*, 57, 6999, 1992.
21. Dirlikov, S. K., Propargyl-terminated resins — a hydrophobic substitute for epoxy resins, *High Performance Polym.*, 2, 67, 1990.
22. Barrall, E. M. and Johnson, J., A review of the status of polymerization in thermotropic liquid crystal media and liquid crystalline monomers, *J. Macromol. Sci. Rev. Macromol. Chem.*, C17, 137, 1979.
23. Paleos, C. M., Polymerization in organized systems, *Chem. Soc. Rev.*, 14, 45, 1985.
24. Douglas, W. E. and Overend, A. S., Curing reactions in acetylene terminated resins. I. Uncatalyzed cure of arypropargyl ether terminated monomers, *Eur. Polym. J.*, 27, 1279, 1991.
25. Benicewicz, B. C., Douglas, E. P., and Hjelm, R. P., Molecular composites from liquid crystalline polymers and liquid crystalline thermosets, in *Liquid Crystalline Polymers*, Carfagna, C., Ed., Pergamon Press, New York, 1994, 87.

26. Hjelm, R. P., Douglas, E. P., and Benicewicz, B. C., The solution structure of a liquid crystal polymer with small liquid crystal thermosets, *Polym. Mater. Sci. Eng.*, 71, no. 2, 287, 1994.
27. Hwang, W.-F., Wiff, D. R., Benner, C. L., and Helminiak, T. E., Composites on a molecular level: phase relationships, processing, and properties, *J. Macromol. Sci. Phys.*, B22, 231, 1983.
28. Krause, S. J. and Hwang, W., Recent advances in morphology and mechanical properties of rigid-rod molecular composites, *Mater. Res. Soc. Symp. Proc.*, 171, 131, 1990.
29. Pawlikowski, G. T., Dutta, D., and Weiss, R. A., Molecular composites and self-reinforced liquid crystalline polymer blends, *Annu. Rev. Mater. Sci.*, 21, 159, 1991.
30. Flory, P. J., Statistical thermodynamics of mixtures of rodlike particles. V. Mixtures with random coils, *Macromolecules*, 11, 1138, 1978.
31. Abe, A. and Flory, P. J., Statistical thermodynamics of mixtures of rodlike particles. II. Ternary systems, *Macromolecules*, 11, 1122, 1978.

Part II

Polyimides and Electronics

Chapter 6

POLYIMIDE SYNTHESIS AND INDUSTRIAL APPLICATIONS

Cyrus E. Sroog

CONTENTS

I. Abbreviations 100

II. Introduction 100

III. Polyimide Synthesis... 100
   A. Two-Step Procedure 100
   B. One-Step Procedure... 101
   C. Vapor Deposition/Polymerization of
      Aromatic Monomers.. 102

IV. Polyimide Properties 103
   A. Thermal Stability 103
   B. Hydrolyic Stability 105
   C. Structural Properties 105

V. Melt-Processible Polyimides 106
   A. Mitsui–Toatsu New-TPI 108
   B. NASA LARC-IA 111

VI. Polyimides with Liquid Crystal (LC) Properties 113
   A. LC Polyimide Copolymers... 113
   B. Blends of Polyimide with Non-Imide LC Polymers.... 116

VII. Polyimide Fibers 116

References 121

0-8493-7672-6/97/$0.00+$.50

99

I. ABBREVIATIONS

PMDA	pyromellitic dianhydride
ODA, POP	4,4'-oxydianiline
BPDA	biphenyl dianhydride
PFMB	2,2'-di-bis(trifluoromethyl)benzidine
DPPMDA	3,6-diphenylpyromellitic dianhydride
PPD	p-phenylene diamine
MPD	m-phenylene diamine
T	terephthalic acid
I	isophthalic acid
TMA	trimellitic anhydride
10	didecanoic acid
PI	polyimide
AI	amide-imide
6F	4,4'-hexafluoroisopropylidene bisphthalic anhydride
ODPA	oxydiphthalic anhydride
BTDA	3,3',4,4'-benzophenone tetracarboxylic dianhydride
Bz	benzidine (4,4'-diaminobiphenyl)
DPE	diphenyl ether

II. INTRODUCTION

This chapter is a review of the present state of high-performance polymers -- science and practice. Principal attention will be given to polyimides with reference also to other high-performance polymers.

Polyimide chemistry has broadened phenomenally in the last decade covering a multiplicity of chemical structures and broad and diverse industrial outlets. Discussion includes unusual forms of synthesis and a variety of structural parameters relating to crystalline/amorphous transitions, solubility, moldability, melt extrudability, and color. Polyimides with liquid crystalline (LC) structures are reviewed as well as the effect on polyimide flow and other properties which result from blended LC polymers.

III. POLYIMIDE SYNTHESIS

A. Two-Step Procedure

The two-step technique (see Table 1) appeared in the 1950s and was the basis for initial development of the aromatic polyimides.[1-3] There are two products formed from the amic acid precursor, namely, the polyimide and also the isomeric polyisoimide. The isoimide forms in significant quantities when the conversion proceeds with catalysis by strongly acidic systems such as PCl or trifluoroacetic anhydride, but only in limited quantities thermally or by the use of such agents as acetic anhydride/pyridine.

TABLE 1.

Two-Step Synthesis of Polyimides[1-3]

Two-step: dianydride + diamine

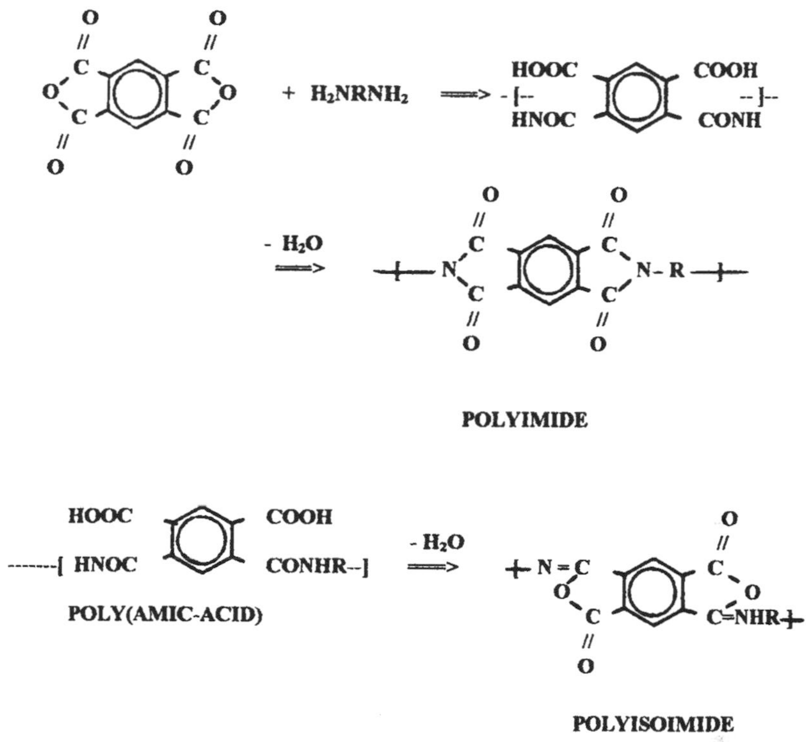

POLYIMIDE

POLY(AMIC-ACID)

POLYISOIMIDE

Polyisoimide converts to normal imide at temperatures above 300°C. This chemistry is the basis for such films as Kapton and Upilex S, molding polymers such as Vespel, and the variety of end uses including coatings of chips and wire.

B. One-Step Procedure

Direct polymerization to polyimide is described in Table 2 which illustrates the reaction and references.[4-8] The invention by Vinogradova et. al.[4] involved direct polymerization of diamine/dianhydride in phenols at high temperature in the presence of tertiary amine catalysis. Later work at Ube[5,6] eliminated the use of amine catalysis, and a large variety of polyimides previously difficult to prepare or completely unavailable were synthesized. More recently, Harris et al.[8] at the University of Akron demonstrated that refluxing N-methylpyrrolidinone (NMP) could be a solvent permitting direct preparation of polyimides soluble at high temperature. Among the polymers successfully prepared were those from biphenyl

TABLE 2.

One-Step Synthesis of Polyimides[4–8]

One-step: dianhydride + diamine in high-boiling solvent
  High-boiling phenols or NMP
  Direct formation of polyimide in solution
  ODA-PMDA and PPD-BPDA are insoluble under these conditions but
    large number of polyimides can be prepared this way,
      e.g., ODA-BPDA
            PFMB-DPPMDA
  Particularly useful for monomers reluctant to polymerize via amic acid route

---

dianhydride (BPDA) with 4,4'-oxydianiline (ODA) and a variety of BP-
DA/PMDA copolymers. These two synthesis procedures make it possible
to prepare and explore a great variety of structures, many of which have
been previously unavailable.

C. Vapor Deposition/Polymerization of Aromatic Monomers

This novel chemistry (see Tables 3–6) proceeds by appropriate monomers (e.g., ODA/PMDA) in a process of vaporization in separate aerosol
type streams which are at high temperature and are carried in nitrogen;[9–16]
these are joined at a hot target (i.e., metal, ceramic polyimide film) with
surface temperatures as high as 300°C. The result is a set of films which
show significant property differences when compared with those of films
conventionally prepared. These include higher density and lower water
absorption and permeation of oxygen or water vapor; there are also indications of reduction of coefficients of thermal expansion. This chemistry
should be studied fundamentally to determine those key structural components contributing to the property differences, and also in practical
terms to explore the range of product possibilities. In addition to polyimides, polyamides and poly(amide/imides) were also examined showing
similar results and thereby indicating the breadth of possibilities.

TABLE 3.

Vapor-Phase Polyimide Synthesis[9–16]

Angelo et al. (DuPont, 1978–79)
- Polyimide films prepared and deposited via high-temperature vapor-
  phase reaction resulted in coated substrates of sharply reduced gas
  permeability, to levels significantly lower than conventionally
  prepared film or coating

Numata et al. (Hitachi, 1988)
- Vapor deposited polyimide from AB monomer (e.g., 4-aminophthalic
  acid ester), reference made to reduced water vapor permeability and
  absorption, and reduced coefficient of thermal expansion (CTE) vs.
  conventionally prepared polyimide film

TABLE 4.

Vapor Deposition/Polymerization

- Can proceed under vacuum or atmospheric pressure
- Target surface needs to be at a minimum coalescence temperature
- Broad variety of polymer types have been demonstrated
- Unique reduction of gas permeation vs. conventional
- Reduction of water vapor absorption and permeation vs. conventional
- Increase of density vs. conventional

TABLE 5.

Minimum Coalescence Temperatures

VPP Polymer System	Coalescence Temp (°C)
PPD-T polyamide	160–170
MPD-I polyamide	142–147
PPD/TMA polyamide-imide	170–180
POP/TMA polyamide-imide	145–152
PPD/PMDA polyimide	~130
ODA/PMDA polyimide	159–165

TABLE 6.

Oxygen Barrier Comparison: VPP vs. Conventional

	VPP I.V. (H_2SO_4)	OPV[a] VPP	Conv.
Polyamide			
PPD-T	2.0-3.0	0.004	0.6
MPD-I	0.3-0.5	0.16	2.5
MPD-10	1.0	4.4	4.4
Polyimide			
PPD-PI	Insol.	0.3	0.9
ODA-PI	0.3-0.5	1.0	7.4
Polyamide-imide			
PPD/TMA-AI	0.3	0.01	0.07

[a] Units: cc/100 in² - 24 hr. - atm

IV. POLYIMIDE PROPERTIES

A. Thermal Stability

The initial impetus to polyimide development was based on the outstanding thermal durability of the aromatic polyimides, as shown in Tables 7–9.[17–20] Considerable research was conducted on thermal stability

TABLE 7.

Properties of Polyimides: Thermal Stability[17-20]

- All-aromatic systems are superior to polyimides which contain aliphatic structures.
- Simple weight loss data are inadequate to predict utility since such effects as cross-linking/embrittlement are not picked up in TGA curve.

For all-aromatic systems:
  Dianhydrides
    PMDA, BPDA, ODPA BTDA are roughly equivalent
    6FDA ~ PMDA
  Diamines
    PPD ~ MPD ~ benzidine > ODA
    2,2'-dichlorobenzidine ~ benzidine

TABLE 8.

Thermal Stability of Polypyromellitimide Films in Air[3]

R (DIAMINE)	275 degC	300 degC
 (phenylene)	> 1 YR.	
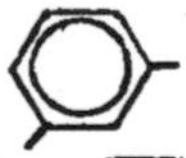 (phenylene, ortho)	> 1 YR.	> 1 MONTH
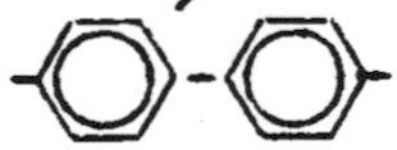 (biphenyl)		1 MONTH
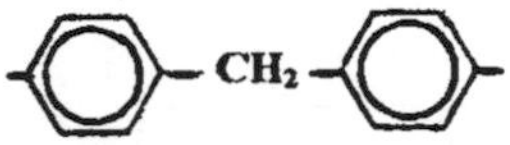 —CH₂— linked diphenyl		7 -10 DAYS
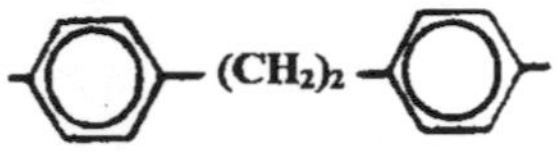 —(CH₂)₂— linked diphenyl		15-20 DAYS
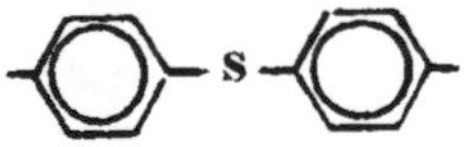 —S— linked diphenyl	CA. 10-12 MONTHS	6 WEEKS
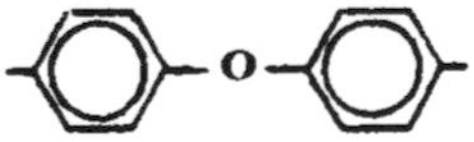 —O— linked diphenyl	> 1 YR.	> 1 MONTH
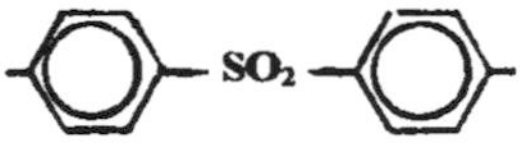 —SO₂— linked diphenyl		> 1 MONTH

in air and nitrogen and significant generalizations were developed. It was recognized very early, however, that utility required a set of additional properties including such key characteristics as excellent mechanical and electrical properties as well as hydrolytic stability and resistance to imbibition of water and gases. The best example is ODA-pyromellitic dianhydride (PMDA), widely known as the structure of Kapton, which had significantly lower thermal stability than polyimides derived from m-phenylene diamine

TABLE 9.

Polyimide Oxidative Stability: Hours to Failure in
Air at 300°C[17]

Polyimide	Hours to Failure	Initial % Elongation
ODA-BPDA	4100	87
ODA-PMDA	2150	70
ODA-ODPA	2900	85
PPD-ODPA	3700	59
Bz-BPDA	2900	30

(MPD) or *p*-phenylene diamine (PPD), but nevertheless had the benefit of other supportive properties not available in the MPD or PPD systems.

B. Hydrolyic Stability

The invention of polyimides brought with it the realization that many otherwise useful polyimides had poor stability under exposure to aqueous caustic or to water alone. In recent years it has become evident that some polyimides can exhibit excellent hydrolytic stability; these involve multi-ring structures either diamine or dianhydride with the juncture proceeding directly (e.g., BPDA) as in Upilex S or through a separate structure (e.g., ODA). See Tables 10–12 and References 21 and 22. For many years these differences were thought of in structural terms of the imide itself, but Kreuz[21] has conducted research relating to amic acids (Table 10). A variety of poly(amic acids) were exposed to aliquots of water added to the DMAC solutions. Excepting some anomalous results (i.e., high rate of molecular weight loss for ODA-oxydiphthalic anhydride[ODPA]), there was surprising similarity in rates of degradation. These results are very preliminary and will clearly require additional research; the implications, however, are intriguing since the possibility exists that hydrolytic stability could relate to completeness of ring closure (i.e., degree of polyimide conversion) rather than the nature of the polyimide.

C. Structural Properties

A great deal of research has been devoted to defining the relationship of structure to such properties as glass transition temperature (T_g), crystalline melting point (T_m), solubility, color, and moldability.[17,18,23-27] Not surprisingly, as shown in Tables 13–15, the data show that chemical structure (e.g., see Kapton vs. Upilex S, Table 13) has an enormous impact on the properties in a manner reasonably predictable from polymer theory, with the exception of polyimide color. Thus, increased internal mobility will decrease crystallinity, increase solubility, and increase moldability. In the case of aliphatic substitution, it appears that crystallinity will decrease

TABLE 10.

Polyimide Hyrolytic Stability[21]

- Strongly dependent on specific structures
- Rigid systems tend to be less stable than flexible systems

    e.g., PPD-PMDA — poor

    PPD-BPDA — good

    PPD-ODPA — good

- Kreuz has done excellent work on hydrolytic degradation of poly(amic acids) of varying structure which could support the belief that incomplete conversion of polyimide could be major source of hydrolyic instability of final polyimide
- Additional research needed to develop broad understanding related to breadth of polyimide structures

TABLE 11.

Hydrolytic Stability of ODPA-Derived Polyimide Films[22]

	PPD-ODPA	PPD-BTDA
Tensile strength, kpsi		
Start	8.5	9.3
Finish	7.5	0[a]
% Retained	88	0
Tensile modulus, kpsi		
Start	29	15
Finish	20	0[a]
% Retained	69	0

Note: Chemical resistance: 14 d in 100°C hot water.
[a] Disintegrated.

without decrease of T_g. Of interest is the high T_g of polyimides relative to their T_m, in considerable contrast to polyesters where the ratios of T_m to T_g in degrees Kelvin is of the order of 3:2. Note should also be made of the change of color with chemical structure. A simple decrease in mobility is not enough to reduce color. Reduction of color also involves introduction of electron withdrawing groups as determined in the pioneering work of St. Clair and Slemp[24] and St. Clair and St. Clair,[26] as shown in Tables 16 and 17. It should be noted that colors of useful polyimides now range from deep yellow-orange to colorless.

V. MELT-PROCESSIBLE POLYIMIDES

There have been long-time indications that polyimides could be melt processible. Prior to the invention of aromatic polyimides, it was known that certain polypyromellitimides could be melt extruded; thus

TABLE 12.

Hydrolyic Stability of OPDA-Derived Polyimide Films[22]

	Polyimides with 4,4'-ODA			
	ODPA	**PMDA**	**BPDA**	**BTDA**
Tensile strength, kpsi				
Start	20	15	21	22
Finish	18	0[a]	25	24
% Retained	90	0	119	109
Tensile modulus, kpsi				
Start	525	480	500	600
Finish	525	0[a]	580	610
% Retained	100	0	116	102

Note: Chemical resistance: 5 d in 10% aqueous NaOH.

[a] Disintegrated.

TABLE 13.

Comparison of Polyimide Film Properties[23]

Property	Kapton Type H (ODA-PMDA)	Upilex Type S (PPD-BPDA)
Tensile strength, MPa		
25°C	172	276
200°C	117	151 (300°C)
% Elongation,		
25°C	70	30
200°C	90	30
Tensile modulus, GPa		
25°C	3.0	6.2
200°C	1.8	6.2
CTE	2.0×10^{-5} (in./in. 1°C)	0.8×10^{-5} (cm/cm 1°C)
Moisture regain, %	1.3 (50% RH)	0.9 (60% RH)
T_g, °C	385	>500

TABLE 14.

Polyimide Structure vs. Thermal Transitions[17,18]

Structure	Crystalline (T_m)	Amorphous (T_g)
Complete aromaticity	+	+
Aromatic with bridged ring	−	−
Alkyl substitution	−	− or +
O or S bridging	−	−
o, p' Catenation	−	−

the polypyromellitimide from nonamethylene diamine or from the 4,4'-dimethylheptamethylene diamine could be melt processed. Successful melt processing of aromatic polyimides has, however, been achieved only

TABLE 15.

Polyimide Structure vs. Solubility and Moldability[18]

Structure	Solubility	Moldability
Complete aromaticity	−	−
Bridging groups diamine O, S, SO$_2$	+	+
o, p' Catenation	+	+
p, p' Catenation	0	0
Cycloaliphatic	+	+
Nonimide copolymerization	Varies with system	+
Plasticization	N.a.	+

TABLE 16.

Reduction of Color in Polyimides[24,25]

Factors which favor a decrease of color

- Bulky electron-withdrawing groups
  e.g., SO$_2$, C(CF$_3$)$_2$
- Separator groups
  e.g., O, S
- Interruption of symmetry
  m, m' ; o, p'

recently. Gannett et al.[27] and Gannett and Gibbs[28] showed that diamines derived from phenoxybenzene resulted in melt-formable polyimides if prepared by unbalanced stoichiometry (see Table 18).

A. Mitsui–Toatsu New-TPI

Mitsui–Toatsu[29-31] conducted research described in patents issued in 1991 on melt-processible polyimides designated as New-TPI. See Tables 19 and 20. New-TPI is prepared by conventional two-step methods in DMAC, but with unbalanced stoichiometry; the polymer as prepared has excess diamine and is end capped with phthalic anhydride. The resulting polymer after polymerization is precipitated and dried before grinding into powder suitable for melt extrusion; melt extrusion proceeds at 390 to 400°C in extruders made of special steels. Thus far, Mitsui–Toatsu has described three molecular weights without specific identification of inherent viscosities (see Figure 1). Careful examination of Figure 1 indicates that these three materials show very little decrease of melt viscosity with increase of shear rate. Poly(ethylene terephthalate) at the shear rate 500 to 1000 s^{-1} used for extrusion into film has a melt viscosity of 1500 to 2000 P. This value is considerably lower than the lowest values for New-TPI. It would appear, therefore, that even the lowest molecular weight

TABLE 17.

Polyimide Color vs. Structure[26,27]

Dianhydride	Diamine	Color
6F	m,m'-SO$_2$D	Colorless
	4-BDAF	Colorless to pale yellow
	p,p'-DABP	Pale yellow
	m,m'-DABP	Pale yellow
	p,p'-MDA	Yellow
	m,m'-MDA	Yellow
	ODA	Pale yellow
	m,m'-ODA	Essentially colorless
	m,p'-ODA	Pale yellow
	APB	Pale yellow to colorless
PMDA	m,m'-SO$_2$D	Yellow
	4-BDAF	Orange-yellow
	p,p'-DABP	Yellow
	p,p'-MDA	Yellow
	ODA	Yellow
	PPD	Yellow
	APB	Yellow
ODPA	4-BDAF	Pale yellow to colorless
	m,m'-DABP	Yellow
	p,p'-DABP	Light yellow
	ODA	Yellow
	PPD	Yellow
	m,m'-ODA	Essentially colorless
	m,p'-DABP	Yellow
	m,p'-ODA	Essentially colorless
	m,m'-MDA	Dark yellow
	APB	Pale yellow

TABLE 18.

Melt-Fusible Aromatic Copolyimides from PMDA[27,28]

1,3-bis(3-Aminophenoxy)benzene (75 mol%)/4,4'-diaminodiphenylether (25 mol%)
1,3-bis(3-Aminophenoxy)benzene (80 mol%)/m-phenylene diamine (20 mol%)
1,3-bis(3-Aminophenoxy)benzene (80 mol%)/p-phenylene diamine (20 mol%)
2-Phenyl-1,4-bis(4-aminophenoxy)benzene (50 mole%)/4,4'-diaminodiphenylether (50 mol%)

New-TPI might be difficult to extrude into thin film of 25 μm or less. Table 21 summarizes the properties of New-TPI Regulus film.[32] The data which mostly show generally desirable properties include two relatively low values which conceivably could be related to low molecular weight: these are the 30% elongation for oriented Regulus film vs. the Kapton value of 70%, and the low values of MIT Flex of 440 and 2,400 for

TABLE 19.

Mitsui–Toatsu: New-TPI[29-31]

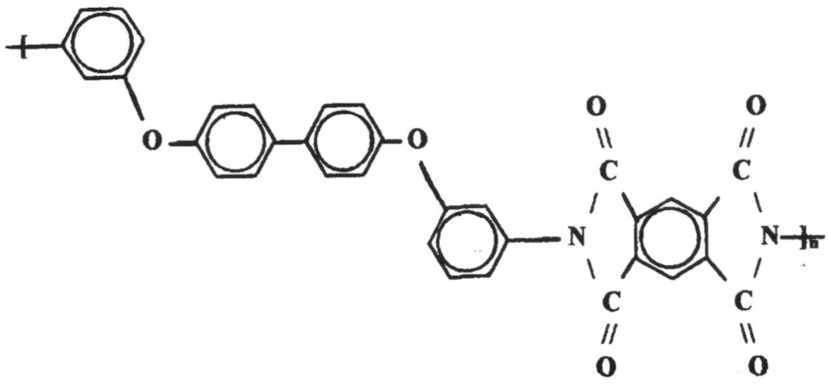

Note: Melt processible at 400°C; T_g = 250°C, T_m = 388°C.

TABLE 20.

Extrusion Guidelines: New-TPI

Type of extruder
  New-TPI can be melted and pumped to a die using a conventional single screw
  extruder (the following points, however, may be noted)

Material
  Metals having as low oxidative degradation as possible should be used as surface
  materials in contact with melt polymers

Screw
  L/D = 20–30, CR = 1.3–1.6 for standard type amorphous pellets, or 2.2–3.5 for annealed
  crystalline pellets

Temp control:
  Extruders equipped with forced air cooling by blowers as well as heating by bronze
  heater can be used, and the temperature may as well be controlled by using PID
  controller

Drive system
  AC or DC motor may be used

unoriented and oriented film vs. a value of 10,000 for Kapton. The latter
may indicate there could be long-term durability problems for New-TPI
film, although thus far there have been no reports of poor durability.
The remaining oriented film values of tensile strength, tensile modulus,
electrical properties, and water absorption are equal to or superior to
Kapton. New-TPI has begun penetration of high-performance polymer
molding applications: the U.S. company Furon-Dixon, Inc. (Bristol,

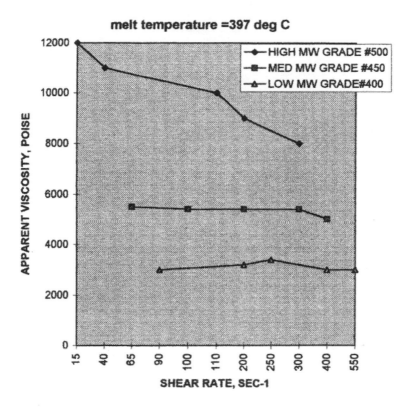

FIGURE 1
New-TPI melt viscosity.

Rhode Island) provides filled and unfilled New-TPI molding resin. Table 22 shows data kindly provided by Furon-Dixon which compares the New-TPI neat resin with Vespel, PEEK, and Torlon. There seems little doubt that the molded New-TPI has a useful and respectable set of properties, with no evidence for possible poor durability noted previously for Regulus film. New-TPI resin is currently being marketed by Advanced Web Products (Ingomar, Pennsylvania) under the trade name Aurum for a variety of high-temperature molding and film applications, including electronics and precision wear-resistant parts in machinery, automotive and aerospace.

B. NASA LARC-IA

NASA has patented LARC-IA,[33,34] a new polyimide from monomers 3,4'-ODA and ODPA. This polymer, shown in Table 23, is prepared similarly to New-TPI by a two-step solution method utilizing unbalanced stoichiometry with phthalic anhydride end capping resulting in a polymer of about 0.47 inherent viscosity. There have been no summaries of viscosity

TABLE 21.

Properties of New-TPI (Regulus) Film[32]

Property	Unoriented	Oriented
Thermal		
T_g, °F	482	482
T_m, °F	730	730
Mechanical		
Tensile strength, kpsi	17.1	37
Elongation, %	110	30
Tensile modulus, kpsi	441	711
MIT Flex, 50 μm cycles	440	2400
Tear propagation, g/mil	50	—
Light transmission, %	64 (35 μm thickness)	—
Dielectric constant, 1 kHz	3.1	3.1
Dissipation factor, 1 kHz	0.0009	0.0009
Dielectric strength, V/μm	280	280
Water absorption, %	0.24	0.20

TABLE 22.

Comparison of New-TPI Properties with Other Neat Resins

	Thermoplastics		Non- or Semi-Thermoplastics	
Property	NEW-TPI unnannealed	PEEK	Vespel	Torlon
Specific gravity	1.33	1.33	1.43	1.40
Tensile strength, kpsi	13.4	14.1	10.5	27.0
Elongation, %	90	80	7	12
Flexural strength, kpsi	19.9	20.6	13.9	30.7
Flexural modulus, kpsi	384	540	360	664
Izod impact (notched)	1.7	1.3	0.3	2.6
Compressive strength, kpsi	17.4	17.4	3.5	44.0
T_g, °F	482	290	—	—
LOI (3.2 mm), %	47	35	—	43
HDT (18.6 kgf/cm²), °F	460	305	680	533
CTE × 10^{-5}/°F	3.1	2.8	2.8	2.2
Water absorption (23°C/24 h), %	0.34	0.14	1.30	0.28
Dielectric constant, 1 kHz	2.7	3.4	3.6	3.4
Dissipation factor, 1 kHz	0.003	0.004		0.026
Volume resistivity, ohm-cm	10^{17}	10^{16}	10^{16}	10^{16}

(Courtesy: Furon-Dixon, Inc.)

vs. shear rate, but the polymer is clearly melt processable at temperatures well below 400°C. Polymer has been molded into unfilled discs at 300°C while a composite of graphite fiber and coated resin was prepared at 350°C from previously prepared piles of fiber and resin. This is a very important new polymer which is already having major

TABLE 23.

NASA LARC-1A[33-34]

Melt processible at 340°C; T_g = 240°C, T_m = 330°C.

influence. NASA does not supply polymer samples but quantities up to 25 kg can be ordered from the U.S. company, Imitec, Inc., of Schenectady, New York.

Additional polymers with the prospect for melt processibility include NASA poly(imide/ether ketones).[35] Structural summaries for these polymers appear in Table 24. Poly(ether ketones) prepared via Williamson chemistry or via Friedel-Crafts procedures have been known for many years as important melt-extrudable polymers. Data for melt-processible poly(ether ketone) film derived from polymer prepared via Friedel–Crafts catalysis appear in Tables 25–27.[36] These indicate a respectable level of properties but with problems associated with their relatively low T_g. The data for both mechanical and electrical properties indicate serious deterioration above 155°C, thereby emphasizing the need for high T_g levels for both melt-processible polyimides and poly(imide/etherketones).

VI. POLYIMIDES WITH LIQUID CRYSTAL (LC) PROPERTIES

A. LC Polyimide Copolymers

Extensive research has been conducted in LC imide/ester copolymers, with expected advantages summarized in Table 28. Kricheldorf[37] has prepared a series of LC copolymers following earlier work by Irwin,[42] as shown in Table 29. The data establish clearly the LC nature of these copolymers, particularly with reference to the determination of anisotropic/isotropic changes and the type of liquid crystal obtained. Adduci et al.[38] conducted a very interesting study comparing solution polymerization to melt polymerization of LC polymers with the same chemical structures (see Table 30). This research indicated that melt polymerization resulted in higher molecular weight vs. solution polymerization. Of

TABLE 24.

Poly(imide/etherketones)[35]

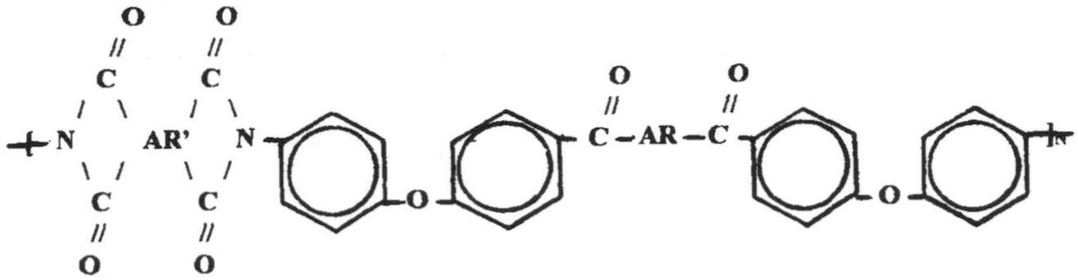

POLYIMIDE	AR'	AR	T_G / T_M (deg C)
PMDA/1,3-BABB			247/ 442
BTDA/1,3-BABB	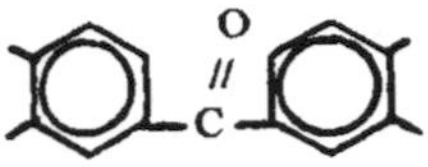	"	222/ 350
BTDA/ 1,4- BABB	"		233/ 427
BTDA/ 4,4' - BABBP	"	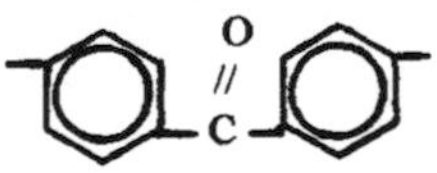	233/ 422
BTDA/ 4,4'-BABDE	"	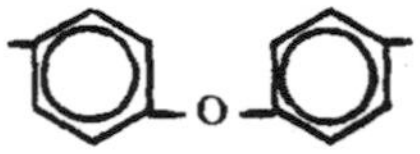	215/ 418
ODPA/ 1,3- BABB			208/ NONE
BDSDA/ 1,3- BABB	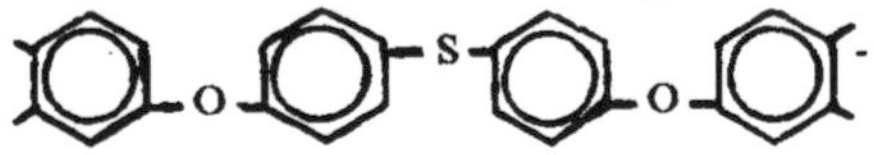	"	192/ NONE

TABLE 25.

Properties vs. T/I Ratio: Copolyketones
Derived from Diphenyl Ether[36]

Mol% T	Mol% I	T_g, °C	T_m, °C
100	0	185	385
90	10	170	371
80	20	165	360
70	30	160	335
60	40	155	310
50	50	155	330
30	70	150	240

TABLE 26.

Electrical Properties of Melt-Extruded DPE-T/I (70/30)[36]

Temp (°C)	Volume Resistivity ohm-cm	Dielectric Constant $10^2/10^5$ Hz	Dissipation Factor $10^2/10^5$ Hz
23	2.2×10^{17}	3.5/3.4	0.003/0.006
105	2.5×10^{15}	3.4/3.3	0.002/0.005
155	4.4×10^{13}	3.4/3.3	0.05/0.01
167	2.8×10^{13}	3.7/3.4	0.05/0.04
180	1.3×10^{13}	5.1/3.6	0.05/0.04
200	5.7×10^{12}	4.4/3.8	0.03/0.04

TABLE 27.

Mechanical Properties of Melt-Extruded DPE-T/I(70/30)[36]

Temp (°C)	Tensile Modulus (kpsi, MD/TD)	Elongation (%)	Tensile Strength kpsi	Pneumatic Impact (kg-cm/mil)	MIT Flexural Cycles MD/TD
23	356/346	166/140	14.0/11.8	2–3	11,800/10,900
105	195/208	257/256	8.6/9.0	—	—
155	36/61	378/284	6.2/5.7	—	—
180	6/4	Off-scale	2.7/2.3	—	—
200	19/12.1	348/285	7.0/3.9	—	—
250	15/14	279/116	3.7/1.3	—	—

TABLE 28.

Polyimides with Liquid Crystal Properties

Property Significance of LC Polymers:
Moisture regain — decrease
Melt or solution viscosity — decrease
Tensile strength — increase
Modulus — increase
% Elongation (reduced significance in blends or copolymers) — decrease
CTE — decrease

considerable practical significance is that Adduci's polymers are nematic in structure and therefore potentially capable of improving melt flow vs. the structures prepared by Kricheldorf which were smectic. Recent work by Kricheldorf et al.[39] has demonstrated the first synthesis of a completely aromatic LC poly(ester/imide) which is nematic to 500°C (see Table 31).

Practical utility of LC poly(imide/esters) is beginning: Teijin has patented a LC copolymer derived from PMDA/trimellitic anhydride derivatives with PPD/ClPPD.[40] Fiber prepared via melt extrusion and orientation exhibited mechanical properties superior to Kevlar (see Table 32).

TABLE 29.

LC Poly(Ester-Imides) from 1,12-Diaminodecane bis-Trimellitimide[37]

| | | T_m (DSC) | | |
Biphenol (R)	T_g (°C)	$T_{m1'}$	T_{m2}	LC Phase
Hydroquinone	—	242	266	232–262
Tetrachlorohydroquinone	117	270	323	335–350

B. Blends of Polyimide with Non-Imide LC Polymers

There has been increased research interest in blends of non-LC polyimides with other LC polymers, as shown in Table 33.[41] A key problem is proper dispersal of the LC component, with considerable uncertainty regarding optimum extent of dispersion of the LC with the matrix polymer. The effect of blended LC Xydar on properties of the matrix polyimide LARC-TPI (polyimide from 3,3'-diaminobenzophenone and 3,3',4,4'-benzophenone tetracarboxylic dianhydride [BTDA]) is shown in Figures 2 and 3. Figure 2 illustrates reduction of melt viscosity from 10 Pa s (nonextrudable) to extrudable levels with 10% Xydar.[46] Figure 3 shows the reduction of coefficient of thermal expansion (CTE) from 35×10^{-6} to 17×10^{-6} (Cu level) with 10% Xydar and to 3×10^{-6} (Si wafer level) with 30% Xydar.[46]

VII. POLYIMIDE FIBERS

Polyimide fibers have received continued attention over many years beginning with the pioneering work of Irwin[42] of DuPont (see Table 34). More recent research by Japanese investigators has built on Irwin's early work, demonstrating new property levels which in many cases duplicate or exceed those for Kevlar. Research by Jinda and Matsuda[44] and by Kaneda et al.[43] are summarized in Table 35. The key to outstanding properties appears to be emphasis on copolymers vs. the earlier study of

TABLE 30.

Properties of Melt-Polymerized Poly(Ester-Imides)[38]

$$+ CO-(CH_2)_n-N \underset{\underset{O}{\overset{\overset{O}{\|}}{C}}{\overset{O}{\overset{\|}{C}}}{\bigcirc} N-(CH_2)_n-CO-O-(CH_2)_m +$$

POLYMER		I.V.	M.P.	T_m	$T_{S-N}{}^a$	$T_i{}^b$
n	m	(dl/g)	(°C)	(°C)	(°C)	(°C)
5	8	0.96	228–230	164	—	232
5	10	0.83	206–208	147	182	211
5	12	0.81	204–206	139	193	210
10	6	0.99	180–182	161	—	184
10	8	1.00	178–179	161	—	185
10	10	0.97	167–170	120	150	173
10	12	0.91	158–160	129	134	161

[a] Smectic to nematic transition.
[b] Temperatures of transition from anisotropic to isotropic melt.

homopolymers, two-stage orientation of the wet-cast fibers followed by (or in some cases accompanied by) very high-temperature treatment in the range of 500 to 550°C. It appears that copolymers are more amenable to orientation and property enhancement than homopolymers. Kaneda established that the oriented polyimide fibers could be superior to Kevlar not only in mechanical properties but also in chemical stress such as boiling water, steam, and thermal stability; the polyimide fibers tested had substantial reduction of water uptake. The polyamide fiber had superior resistance to caustics, but the polyimide fiber did not simply fall apart upon exposure to 10% NaOH; exposure to concentrated sulfuric acid showed polyimide fibers to be superior to Kevlar. There is presently no sound theoretical basis to explain the comparisons noted and property levels attained for the polyimides. Kaneda found no evidence of polyimide LC properties; the need for more understanding continues.

TABLE 31.

Completely Aromatic LC Poly(Ester-Imide)[39]

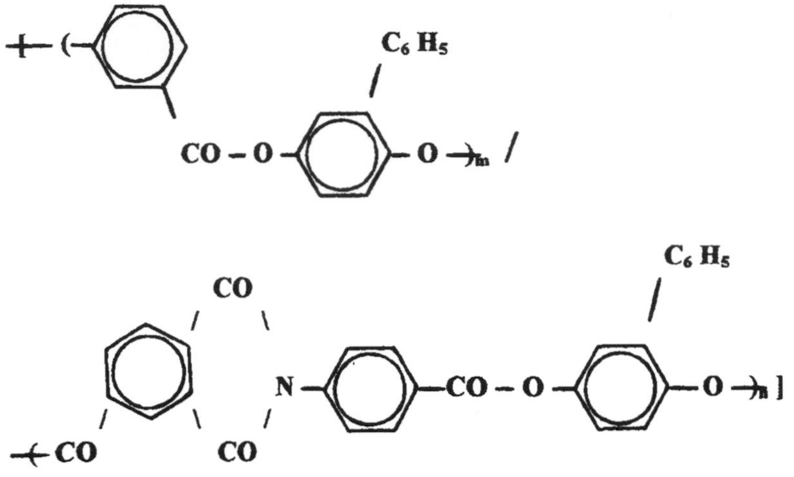

$T_g = 185°C$ for m/n - 1.9; nematic LC.

TABLE 32.

Industrial Activity on LC Poly(Ester-Imides)[40]

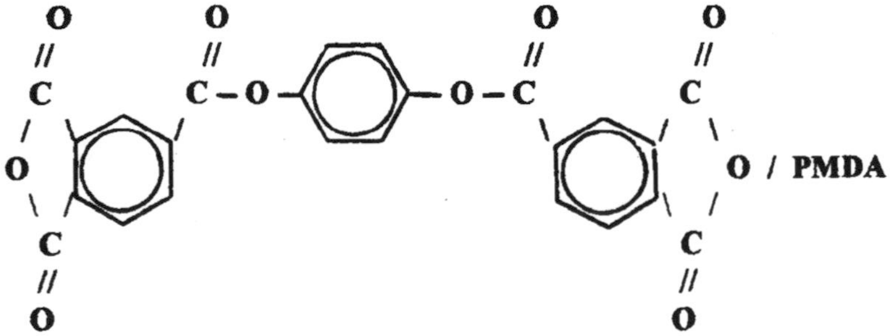

+ CIPPD/PPD $\Rightarrow$ Oriented fibers with properties superior to Kevlar

TABLE 33.

Polyimide Blends with LC Polymers[41]

Blends of thermotropic LC polymers with isotropic substrates:
- PEEK/Ultem/LCP: significant reduction of melt viscosity
- LARC-TPI/Xydar 90/10 (Foster–Miller): major decrease in melt viscosity permitted melt extrusion at 300°C to form clear coherent film

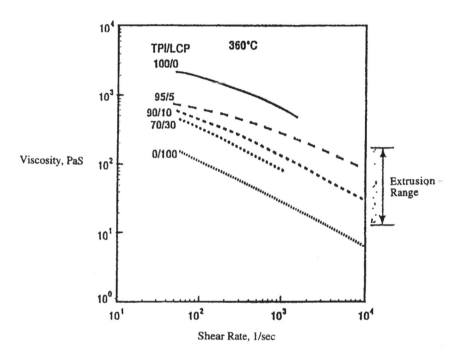

FIGURE 2

Shear viscosity of polyimide blends. (Reproduced with permission of Superex Polymer, Inc.)

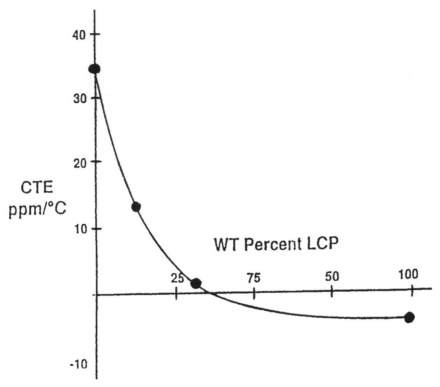

FIGURE 3
Coefficient of thermal expansion vs. weight percent LCP. (Reproduced with permission of Superex Polymer, Inc.)

TABLE 34.

Polyimide Fibers

Irwin (DuPont)[42]

ODA-PMDA
    One-stage draw/550°C treatment
    T/E/M = 0.45 GPa/11.7%/6.4 GPa
PPD/3,4'-ODA-PMDA
    One-stage draw/500°C

Japanese Research: Two-stage draw/500–550°C treatment

Kaneda(Ube)[43]

3,3' DiMeBz-BPDA/PMDA (70–30)
    Superior to Kevlar in thermooxidative stability, resistance to steam and to conc H_2SO_4,
        but poorer in 10% NaOH
    Moisture regain: about 1/6th that of Kevlar 49

Jinda(Toray)[44]

DiCl Bz (1984)

TABLE 35.

Properties of Oriented Polyimide Fibers[45]

Polymer Composition		Draw Ratio (°C)	Tenacity (GPa)	Elongation %	Modulus (GPa)
3,3'-DiMeBz	BPDA/PMDA				
100	100/0	1.6 (500)	1.7	3.0	79
100	80/20	3.2 (500)	3.0	2.4	129
100	70/30	3.8 (500)	3.3	2.6	130
3,4'-ODA	BPDA/PMDA				
	70/30	9.2 (330)	2.4	4.4	63
ClBz/ClPPD	PMDA				
100/0		1.1 (550)	2.0	1.0	205
60/40		1.2 (550)	2.5	1.5	180
DiClBz/ClPPD	PMDA				
70/30		3.9 (550)	3.3	2.0	180
Bz	PMDA/ODPA				
	100/0	1.03 (575)	0.99	1.0	113
	70/30	2.0 (600)	1.5	1.4	129
	40/60	6.5 (550)	2.5	1.7	168
Kevlar 49		—	3.0	2.5	125
Kevlar 149		—	2.3	1.5	142

REFERENCES

1. Endrey, A. L., Canadian Patent, 645,073, 1962.
2. Endrey, A. L. et al., U.S. Patent, 3,179,630–3,179,635, 1965.
3. Sroog, C. E., Endrey, A. L. et al., *J. Polym. Sci.*, A-3, 1373, 1965.
4. Vinogradova, S. V. et al., *Polym. Sci. USSR*, 12, 9, 2254, 1970.
5. Sasaki, Y. et al., U.S. Patent, 4,473,523, 1981.
6. Itatani, H. et al., U.S. Patent, 4,568,715, 1986.
7. Kaneda, T. et al., *J. Appl. Polym. Sci.*, 32,3133, 1986.
8. Harris, F. W. et al., *High Performance Polym.*, 1, 13, 1989.
9. Angelo, R. J. et al., U.S. Patent, 4,104,438, 1978; 4,180,614, 1979.
10. Numata, S. et al., U.S. Patent, 4,759,958, 1988.
11. Ikeda, R. M., Angelo, R. J., et al., *J. Appl. Polym. Sci.*, 25, 1391, 1980.
12. Ijima, M. et al., *Macromolecules*, 22, 2944, 1989.
13. James, S. G. et al., *High Performance Polym.*, 4, no. 1, 3, 1992.
14. Kowalczyk, S. P. et al., *Mater. Res. Symp. Proc.*, 227, 55, 1991.
15. Ijima, M. et al., *J. Appl. Polym. Sci.*, 42, 2197, 1991.
16. Salem, J. R. et al., *J. Vac. Sci. Technol.*, A4, 369, 1986.
17. Bessonov, M. et al., *Polyimides:Thermally Stable Polymers*, Plenum Press, New York, 1987.
18. Sroog, C. E., *Prog. Polym. Sci.*, 16, 561, 1991.
19. Dinehart, R. A. and Wright, W. W., *DieMakro. Chem.*, 153, 237, 1972.

20. St. Claire, T. L., Symp. Recent Advances-Polyimides, Reno, NV, July 1987.21.Kreuz, J. A., *J. Polym. Sci. A*, 28, 3787, 1990.
22. Mundhenke, R. F. and Schwartz, W. T., *High Performance Polym.*, 2, no. 1, 157, 1990.
23. Product Bulletins, Kapton-Type H (DuPont); Upilex S (Ube, Ltd).
24. St. Clair, A. K. and Slemp, W. S., *SAMPE J.*, 28, July 8, 1985.
25. Hogahi, K. and Noda, Y., U.K. Patent, 2,174,399A, 1986.
26. St. Claire, A. K. and St. Claire, T. L., U.S. Patent, 4,603,061, 1986.
27. Gannett, T. P. et al., U.S. Patent, 4,725,642, 1988; 4,485,140, 1984.
28. Gannett, T. P. and Gibbs, H. H., U.S. Patent, 4,576,857, 1986.
29. Misui–Toatsu New-TPI Product Bulletin, 1991.
30. Saruwatari, M. et al., U.S. Patent, 5,037,587, 1991.
31. Saruwatari, M. et al., U.S. Patent, 5,069,848, 1992.
32. Sagita, Y. to Sroog, C. E., personal communication, August 14, 1992.
33. NASA Tech. Briefs, September 1991.
34. St. Claire, T. L. and Progar, D. J., U.S. Patent, 5,147,966, 1992.
35. Hergenrother, P. M. et a.l, *J. Polym. Sci. A*, 25, 1093, 1987.
36. Sroog, C. E., IUPAC-Kyoto, Japan, October 29–November 1, 1985; Abstracts, p. 85.
37. Kricheldorf, H. R. and Pakull, R., *Polymer*, 28, 1772, 1987.
38. Adduci, J. M. et al., *Polym. Prepr.*, 31, no. 2, 63, 1990.
39. Kricheldorf, H. R. et al., *J. Polym. Sci. A*, 31, 279, 1993.
40. Kanda, T. and Matsuda, T., Japanese Patent 63,230,739, 1988.
41. Blizard, K. G. et al., Presentation Ultralloy '90 Houston, TX, November 28–30, 1990.
42. Irwin, R. S., U.S. Patent, 3,415,782, 1968; 4,640,972, 1987.
43. Kaneda, T. et al., *J. Appl. Polym. Sci.*, 32, 3133, 1986.
44. Jinda, T. and Matsuda, T., *Sen Gakkaishi*, 40, 12, 1984; 42, 10, T554, 1986.
45. Kevlar 49 and 149 Data, Kevlar Product Bulletins K3, E-95612.
46. Courtesy Rubin, L. and Lusignea, R., Foster-Miller, Inc.

Chapter 7

ELECTRONIC APPLICATIONS OF HIGH TEMPERATURE POLYMERS

R. Bruce Prime and Claudius Feger

CONTENTS

I.   Introduction 124

II. Storage Technology.. 124
     A.   Magnetic Storage 124
     B.   Optical Storage 125

III. Computer Electronics 125
     A.   Printed Circuit Boards.. 126
     B.   Semiconductors 129
          1.   Lithographic Technology Trends 129
          2.   Bipolar and CMOS Technology Trends 130
          3.   Electronic Packaging 131
          4.   Integrated Circuit Interconnects...................................... 141

IV. Optoelectronic Applications.. 142

V.   Electrically Conducting Polymers 143

VI. Conclusions 144

0-8493-7672-6/97/$0.00+$.50
© 1997 by CRC Press, Inc.

Acknowledgments 144

References 144

I. INTRODUCTION

High temperature polymers, typically those with a glass transition temperature (T_g) greater than 150°C,[1-3] are key materials in the computer electronics industry. There are four major areas in the electronics industry that demand high temperature polymers: computer electronics, information storage, displays, and communications. The largest quantity of high temperature stable polymers is used in computer electronic applications such as the manufacture of semiconductor devices, electronic packages, circuit boards, and interconnects. Materials for all electronic applications need to fulfill several rigorous performance requirements to be considered for eventual application: they must be cost effective; they should be known, off-the-shelf materials which are available from more than one source; they need to conform to current environmental regulations, but also should pose no problems with respect to foreseeable future environmental legislation here and abroad. However, because of the steadily decreasing feature size and increasing device density, the volumes of high temperature materials needed in many electronic applications are shrinking.

Because the processing and lifetime requirements for materials used in the electronics industry are very complex, rigorous testing takes enormous amounts of time and expense. This cannot be done by material vendors or users alone. Therefore consortia and programs sponsored by industry and government have been founded such as Sematech/MCC, National Display Consortium, National Institute of Science and Technology (NIST) ATP program, and the PA dual-use research program. Furthermore, academia is asked to increase the commercial relevance of its research.

II. STORAGE TECHNOLOGY

A. Magnetic Storage

The most stringent requirement for polymers used in magnetic storage devices is the necessity to be noncontaminating. To be avoided are gaseous or volatile organic products which can condense on the magnetic heads or disks, materials such as additives and mold release which can be transferred by contact with critical components, and parts such as machined plastics which can generate particulate contamination. For example, polydimethylsiloxane (PDMS) is generally avoided in magnetic storage devices because of its propensity to outgas low molecular weight species which can be converted to silica-like materials at the head–disk interface. Great care must be given to the cure of adhesives to minimize their outgassing. Machined plastic parts are not used within the disk drive itself since they pose a particulate contamination hazard. Exceptions are polycarbonate (PC) and polyetherimide (PEI) which can be solvent vapor polished to seal the surface.

Important high temperature polymers used in magnetic storage are polyperfluoroethers which are used as lubricants, polycarbonate, polyphenylenesulfide (PPS), PEI, polyimide (PI) films, and aramid fibers. As examples, trays made of PPS are used to carry magnetic heads through a cleaning step in an organic solvent at 80°C, and actuators in which the normally metallic combs are made from PPS are available in the marketplace. PEIs are used widely in structural applications because of their high modulus and their resistance to abrasion. PIs exhibit the same qualities and are widely used in flexible cables. Aramid fibers and poly(tetrafluoroethylene) [PTFE] are used as additives to impart toughness and wear resistance, respectively. Providing the cost is right, one can envision more of the magnetic disk drive being made from high temperature polymers, even the substrate for the disk.

B. Optical Storage

The most widely used polymer in optical storage applications is polycarbonate (PC) (see Figure 1), from which disk substrates are manufactured. It is characterized by low cost, excellent thermal properties, good transparency, and low water absorption. However, the PC chains are prone to align with respect to a substrate or a shear direction causing directional variations in the refractive index. Resulting in-plane birefringence degrades Kerr rotation signals, resulting in loss of signal intensity. Vertical birefringence can introduce a focus effect which produces ambiguity in the best focus position for tracking and data collection. Advances in optical storage technology will lead to increased data density through the use of shorter wavelength lasers and other developments. For this to happen materials and processes for substrate technology need to advance to provide lower birefringence.[4,5] At some point it may be necessary to abandon polycarbonate and move to new materials to meet the more stringent optical and mechanical requirements of future products. One material with much promise is amorphous polyolefin (APO) which does not align during injection molding. However, APO is currently about ten times more expensive than PC.

III. COMPUTER ELECTRONICS

Computers have two main components, integrated circuit logic and storage chips, and the electronic package which encompasses the bulk of the computer (Figure 2).[6] The package is divided into levels such as the

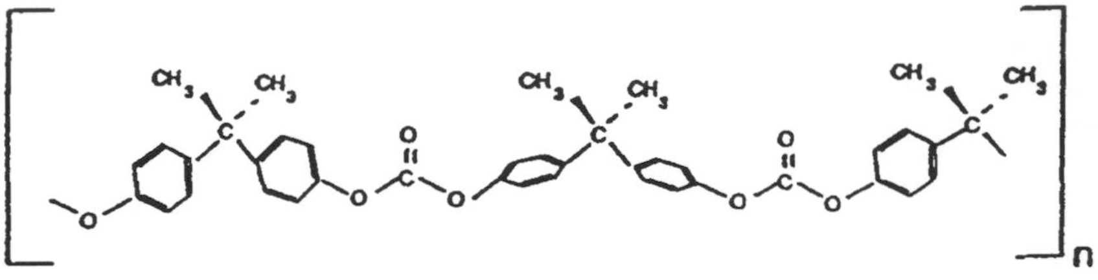

FIGURE 1
Polycarbonate molecular structure.

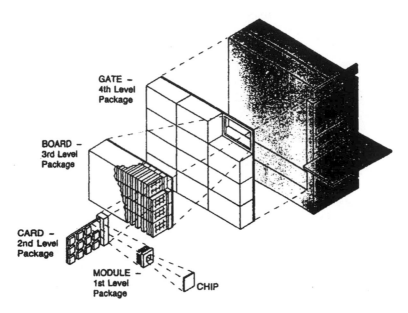

FIGURE 2
Schematic of computer packaging hierarchy. (From Tummala, R. R. and Rymaszewski, E. J., Eds., *Microelectronics Packaging Handbook*, Van Nostrand Reinhold, New York, 1989. With permission.)

first-level package which is also called the module, the second level which might be a card or a printed circuit board, etc. High-thermal stability is required of materials used in chips, modules, cards, and boards to survive several high temperature processing steps, particularly during joining of chips to substrates.

A. Printed Circuit Boards

The manufacture of printed circuit boards (PCBs) is shown schematically in Figure 3. The process indicates that many steps are necessary. These include exposure to aggressive media such as plating baths and

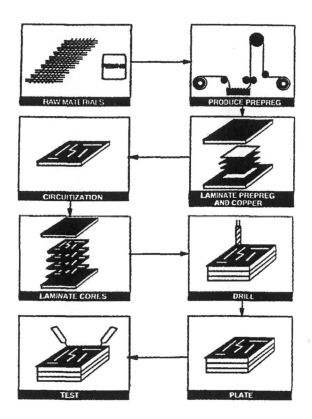

FIGURE 3
Schematic of the PCB manufacturing process.

etching solutions. Others such as drilling of through holes are mechanically demanding. The most important material for PCB manufacture has been FR-4 epoxy (Table 1), which is still widely used in less demanding applications. Better properties, in particular, higher glass transition temperatures, can be obtained with multifunctional epoxies.[7] Many of today's advanced PCBs use polycyanurates. They exhibit excellent thermal stability (decomposition temperatures in excess of 400°C), high glass transition temperatures which can be tailored through changes in the resin chemistry, low dielectric constants (2.6 to 2.8), and outstanding adhesive properties. Further, they are melt processible, are inherently flame retardant without using bromine, and are commercially available in high purity. However, they tend to be brittle.

In a recent development[8] (Figure 4), tough, ductile engineering thermoplastics were incorporated into polycyanurates to enhance toughness without sacrificing high temperature performance. Care had to be taken to produce thermosetting materials with discrete submicron phases of the thermoplastic uniformly dispersed in the thermosetting matrix. This was achieved by utilizing structure property relationships to design a highly

TABLE 1.

Comparison of Matrix Resins for Printed Circuit Boards (PCB)

Property	Difunctional Epoxy (FR-4)	Multifunctional Epoxy (Driclad™)	Toughened Polycyanurates (Cytuf™)
Glass transition temp, (°C)	125	175	200–315 (Tailorable)
Dielectric constant	3.5–3.8	3.7	2.6–2.7
Thermal expansion, ppm/°C	60–75	60–75	49–58
CTE measurement range, °C	<125	<150	<260
Fracture toughness, J/m²	70–90	40–50	200–700 (Tailorable)
Moisture absorption, % (16 h @ 100°C H₂O)	1.93	0.82	0.42

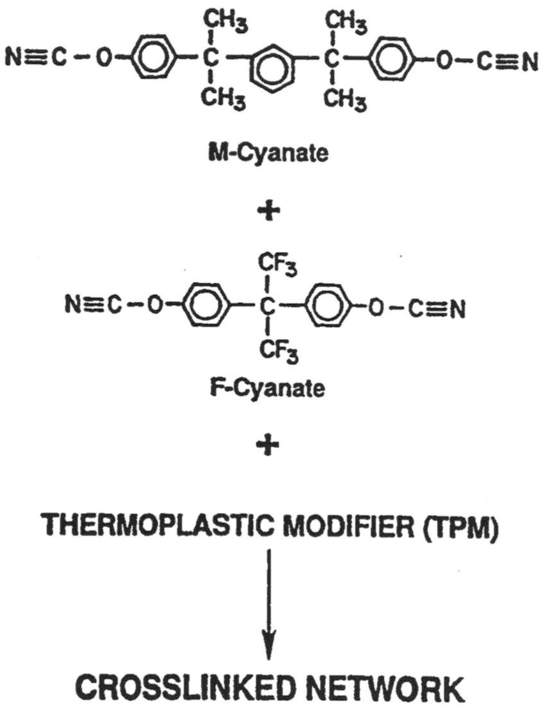

M-Cyanate

+

F-Cyanate

+

THERMOPLASTIC MODIFIER (TPM)

CROSSLINKED NETWORK

FIGURE 4

Thermoplastic toughening of polycyanurates.

compatible thermoplastic which is both miscible in the resin prior to cure and soluble in the processing solvent. The resulting polycyanurates exhibit improved fracture toughness (Table 1), do not show microcracking during drilling and handling processes, and exhibit increased strength in interlaminar and copper adhesion.

B. Semiconductors

Logic chips have seen a relentless increase in transistor density and speed. Concomitant has been a reduction in the price/performance ratio so that today's desktop computer central processing unit (CPU) is comparable to a mid-80s mainframe CPU. Much of this advance has come by decreasing the dimension of line widths. This continuing trend reflects advances in resist technology which for the 64 megabyte (MB) DRAM requires resolution of 0.35 μm using so-called I-line (365-nm) lithography. Each new generation of DRAM requires a new lithographic technology. The next technology (0.25 μm) will be based on deep UV (248 nm) for the manufacture of the 256 MB DRAM. After that, dimensions will likely shrink to 0.18 μm which can be reached by using ArF excimer laser irradiation (193 nm). Such lithographic techniques would allow manufacture of 1 gigabyte (GB) DRAMs. Figure 5 shows the historical and the projected increase in bits per DRAM chip and circuits per logic chip.[9]

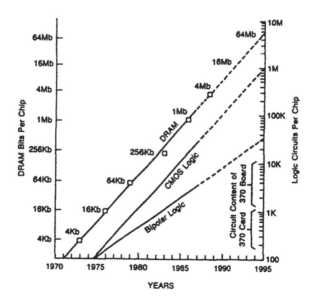

FIGURE 5
Integrated circuit trends. (From Rymaszewski, E. J. and Tummala, R. R., in *Microelectronics Packaging Handbook*, Tummala, R. R. and Rymaszewski, E. J., Eds., Van Nostrand Reinhold, New York, 1989, 1. With permission.)

1. Lithographic Technology Trends

A schematic of the lithographic process is given in Figure 6. Areas of research and development in this area include resists for semiconductors, packaging and magnetic storage, synthesis of new resists, understanding of the photochemistry, and modeling and simulation of lithographic processes.

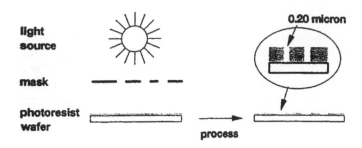

FIGURE 6
The photolithography process.

Typical resists used in current I-line lithographic technology (in use since 1990, 0.75 to 0.35 µm) are based on Novolac resins (T_g = 100 to 110°C) which contain diazonaphtholquinones as the photoactive moiety. The latter decomposes at temperatures of about 125°C. This makes it difficult to evaporate all the solvent before irradiation because solvent diffusion below the glass transition is severely hindered. Outgassing of residual solvent during processing after irradiation can lead to pattern distortion and limits the dimensions achievable. Deep UV resists which will be used from 1995 to about 1999 are based on hydroxystyrene polymers with T_g of 150 to 175°C. The photosensitive component, photoacid generators which lead to acid catalyzed cleavage of the side groups on the backbone, decomposes around 115 to 120°C leading to similar limitations as described above. Deep deep UV (DDUV) resists will drive lithographic technology below 0.25 µm and will be needed from about 1998 onward. Current front-runners are polymethacrylates with T_gs of about 160°C and decomposition temperatures of the photoactive species of about 175°C. The high glass transition temperature which is below the decomposition temperature is key for the bake process in which solvent is removed, stresses are relieved, and adhesion is improved. Also the thermal stability of the image is important. Although the imaging chemistry for 248 and 193 nm technologies is very similar, the polymeric materials used in these processes are completely different.

2. Bipolar and CMOS Technology Trends

Figure 7 shows the trend in power dissipation per chip in watts.[9] The power densities anticipated by this trend are staggering. This has led to a major change in logic chip design. While most high-performance logic chips were being built using fast but power hungry bipolar technology, today's designers have shifted to the slower but lower energy consuming CMOS technology even for high-performance applications. (CMOS logic chips have been standard in desktop and portable computers.) This technology has been used very widely in memory chips because CMOS gates use less space than bipolar gates. A comparison of CMOS and bipolar technologies is given in Table 2.[10] The ramifications of the switch from bipolar to CMOS logic have been profound. Because CMOS chips have higher circuit densities and because CMOS chips tend to be larger than bipolar chips, fewer chip-to-chip interconnects are needed. Fewer chips need fewer packages, in particular fewer multichip modules (MCMs; Figure 8). These have been the focus of intense development efforts in the electronics industry over the past decade.[11] In particular, their use in combination with thin film packages[12,13] (Figure 8) which utilize polyimide/metal composite structures has led to massive development efforts for appropriate high temperature stable polymeric dielectrics. Many of these efforts have slowed down or disappeared altogether because MCMs are only gradually being used to package logic CMOS CPUs.

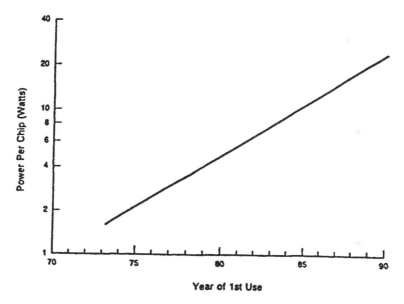

FIGURE 7
Trend in power dissipation per chip in watts. (From Tummala, R. R. and Rymaszewski, E. J., Eds., *Microelectronics Packaging Handbook*, Van Nostrand Reinhold, New York, 1989. With permission.)

3. Electronic Packaging

Because of the central role MCMs have played as a driver for high temperature stable packaging materials and because MCMs will eventually come back as the package of choice for CMOS logic chips, current MCM packages and the polymers developed for applications associated with them will be discussed in more detail.

MCM packages demand high temperature stability[6,11] because most chip-to-MCM package connections are made using controlled collapsed chip connectors (C4s) or solder balls. The reason for this preference is shown in Figure 9[14] which indicates that many more connections per chip

TABLE 2.

Comparison of CMOS and Bipolar Technologies

	1994	2001
CMOS		
(Gate array and MPU)	1.7 (2.0)	2.5
I/O	600 (1000)	2000
Power (watts)	15 (50)	60 (120)
Rise time (ps)	700 (200)	150
Switching lines		
Voltage	3.3 (2.1)	1.6
# Lines (MPU)	128 (256)	400
Bipolar (gate array)		
Size (cm)	2.2	2.5
I/O	600 (1000)	2000
Power (watts)	60 (100)	200
Rise time (ps)	100 (75)	40

From Shambrook, K. P., Proc. of IEEE-CHMT Multichip Module Technologies and Alternatives Seminar, San Diego, CA, April 13, 1991.

NEC SX-3/SX-X Supercomputer

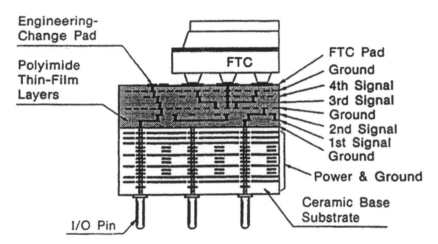

FIGURE 8
Multichip module with thin film package. (After Tummala, R. R. and Rymaszewski, E. J., Eds., *Microelectronics Packaging Handbook*, Van Nostrand Reinhold, New York, 1989. With permission.)

can be made with C4 technology than with other methods. Other driving forces for the development of new organic dielectrics are shown in Figure 10,[15] prominent among them the need to decrease the dielectric constant. Low dielectric constant dielectrics allow increased package signal line

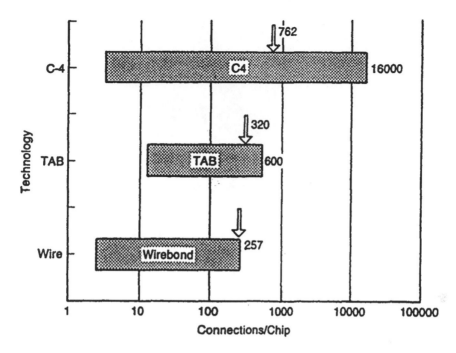

FIGURE 9

Comparison of chip I/O densities achievable using various interconnect technologies. Arrows indicate product usage in 1987. Major improvements increasing I/O densities using wirebonding have been achieved since then. (From Tummala, R. R. and Rymaszewski, E. J., Eds., *Microelectronics Packaging Handbook*, Van Nostrand Reinhold, New York, 1989. With permission.)

densities, higher signal propagation speed, and lower power required for signal switching.[11] Organic polymeric materials allow further lower cost processing than their main competitors, metal oxides and nitrides (SiO_2, Si_3N_4, Al_2O_3, etc.). The major technical challenges for polymers[16,17] in this application is thermal stability; mechanical integrity during processing, in particular stress management; and electrochemical stability which includes corrosion protection and electromigration issues. The ultimate goal for the dielectric in an electronic package is to allow signal transmission with minimal power requirements and maximum signal integrity. This, in turn, requires polymers with uniform dielectric properties (low dielectric constant and low dielectric loss) over a wide range of frequencies. Furthermore, water uptake must be minimal because the latter influences dielectric properties but also leads to corrosion and loss of mechanical integrity (adhesion loss, swelling stresses, etc.).

As pointed out, the most demanding thermal requirements are dictated by the C4 chip joining process which occurs at about 370°C. Temperatures at which metal is deposited can be quite high as well. The mechanical requirements need to ensure that the polymeric material can withstand all processing conditions without mechanical failure such as

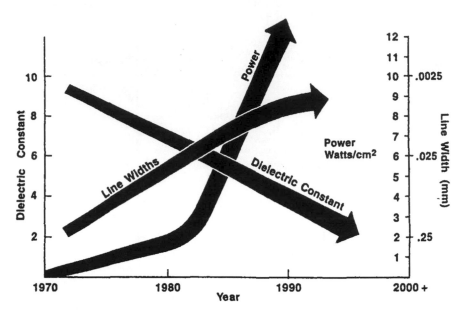

FIGURE 10
Driving forces for the development of new organic dielectrics. (From Rymaszewski, E. J. and Tummala, R. R., in *Microelectronics Packaging Handbook*, Tummala, R. R. and Rymaszewski, E. J., Eds., Van Nostrand Reinhold, New York, 1989, 1. With permission.)

delamination, cracking, or crazing. This is a tall order because most materials with which the polymer is in contact in an electronics package have significantly higher modulus and much lower thermal expansion coefficients so that much stress is exerted on the polymer. Finally, the polymeric dielectric must be compatible with electronics processing.[18] This includes spin apply from solutions which need to have good shelf life, particularly with respect to the solution viscosity, wet etching in caustic media, dry etching by reactive ions in plasmas or by lasers, metal deposition through metal sputtering, evaporation or chemical vapor deposition (CVD), metal plating which involves exposure to plating baths, lift-off processing which involves exposure to hot solvent and ultrasound, and chemical-mechanical polishing to planarize structures which exert chemical and mechanical stresses on the material. During many of these steps the polymer has to be completely inert, should not outgas, should not absorb ions or solvent, etc. From all these requirements a wish list has been extracted which is given in Table 3.

Polymers that have done admirably well in such demanding applications include polyimides.[19,20] The most prominent representatives are the polyimide made from pyromellitic acid dianhydride (PMDA) and oxydianiline (ODA) and, more recently, the one made from biphenyl tetracarboxylic acid dianhydride (BPDA) and *p*-phenylene diamine (PDA). The chemical structures of these and other successful starting

TABLE 3.

Wish List of Properties for the Ideal Polymeric Dielectric

Dielectric constant	2.0 to 3.0 Frequency independent
CTE, ppm/°C	1 to 5
Thermal stability	At 400°C <0.1 wt%/h
Young's modulus	0.7 < × <3.0 GPa
Elongation at break	>10%
Critical crack propagation	Above 2.5% strain
Adhesion	Good to metals, substrate and self
Application	From solution (spin, spray)
Planarization	>90%
Shrinkage	<10 vol%
Water absorption	0.1 to 0.5%
Solvent resistance	No swelling, no crazing
Etchability	RIE, laser, wet: no residue
Color	Light, transparent

From Feger, C. and Feger, C., *Multichip Module Technologies and Alternatives: The Basics*, Van Nostrand Reinhold, New York, 1993, 311. With permission.

materials are given in Tables 4 and 5. Most polyimides are processed from polyamic acid precursors. Cure of the materials involves the loss of solvent and water, and chain alignment and ordering. Chain alignment is at times the cause of very high anisotropy of properties (e.g., BPDA-PDA dielectric constant, out-of-plane: 2.9; in-plane: 3.7[21-23]) while chain ordering and alignment introduce significant process dependence of the final properties (e.g., PMDA-ODA elongation at break: 60% at slow cure; 120% at fast cure[24]). Specific concerns with polyimides from polyamic acid precursors include attack of metals through the acid functionality, poor adhesion to itself, SiO_2 and metals, and often high water uptake. A more expensive but more soluble precursor to polyimides are polyamic acid esters.[25] They often also exhibit better adhesion characteristics. Also acetylene terminated polyisoimides have been used as precursors for dielectric coatings.[26] They exhibit excellent planarization, solubility, and adhesion. However, they tend to show increased swelling in process solvents and often are less thermally stable than traditional polyimides. Also, isoimide solutions are more sensitive toward water contamination.

Of particular interest are photosensitive polyimides (PSPI). These[27-29] materials promise processes which have fewer steps, and thus the overall process has higher yield and lower cost (Figure 11[17]). There are two major types of PSPIs, namely, allyl-containing and inherently photosensitive PSPIs. The former are mostly esters or salts of polyamic acids[27-29] (Figure 12a and b) while the latter contain photosensitive groups such as benzophenone units[30-32] (Figure 12c). Concerns with these materials can include high stress, solvent sensitivity, water uptake, etc.

Possibly the largest use of polyimides in the electronics industry is in the form of films for flex circuits and flex connectors,[33] that is, a flexible polyimide film with metal lines and possibly an attached chip. Nearly

TABLE 4.

Dianhydrides Used in the Synthesis of Electronics Packaging Polyimides

DIANHYDRIDE	STRUCTURE
(PMDA) (P)	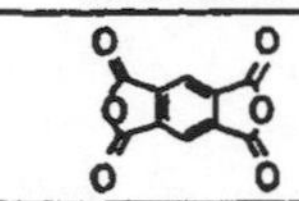
BTDA	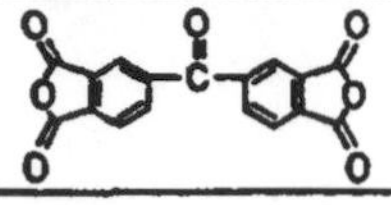
BPDA	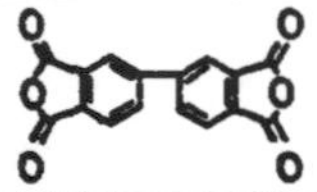
ODPA	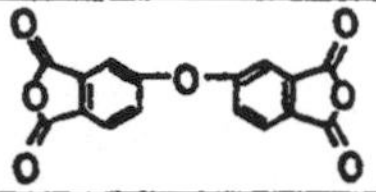
TPDA	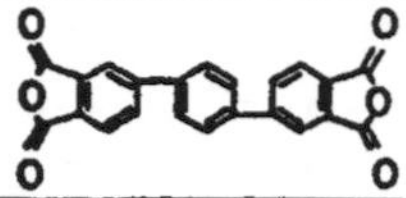
6FDA (HFDA)	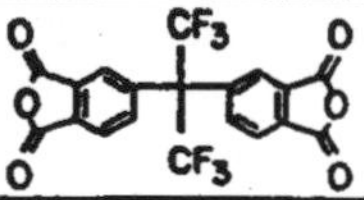

TABLE 5.

Diamines Used in the Synthesis of Electronics Packaging Polyimides

DIAMINE	STRUCTURE
ODA (DDE)	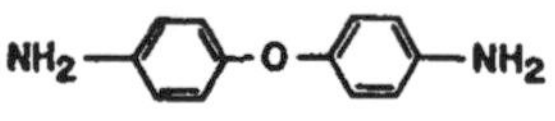
PDA (PPD)	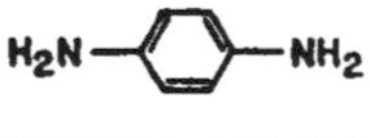
6FDAM (HFDAM)	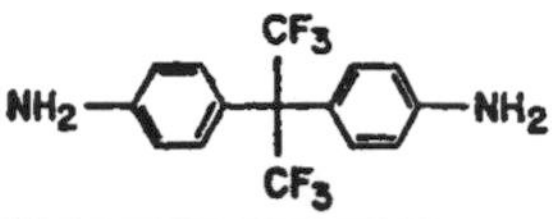
3FDAM	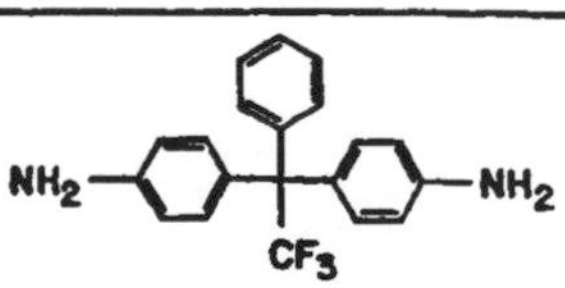

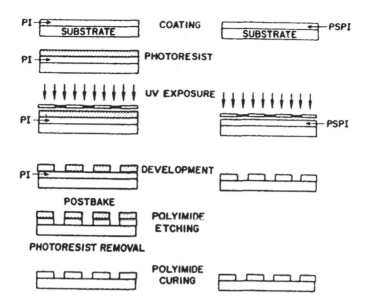

FIGURE 11

Comparison of process steps using conventional and photosensitive polyimide. (From Feger, C. and Feger, C., in *Multichip Module Technologies and Alternatives: The Basics*, Doane, D. A. and Franzon, P. D., Eds., Van Nostrand Reinhold, New York, 1993, 311. With permission.)

every piece of consumer electronics contains a piece of polyimide film in the mentioned capacity. Polyimide films are also being studied for use in high-density interconnects such as MCMs[34,35] (Figure 13). In these processes, metallized and electrically tested interconnection circuits are soldered to each other, a substrate, and chips. Many thousand metal vias need to be connected successfully in the final lamination steps, a daunting task but one that can be done.

Another class of polyimides that has found applications in the electronics industry are siloxane-containing polyimides.[36,37] These materials are either random copolymers or block copolymers. The random copolymers are very similar to polyurethanes in that they have hard and soft segments which may or may not phase separate. Their T_gs vary from 80 to 300°C depending on hard block content. They exhibit excellent adhesion and are used as adhesives but also as chip encapsulants. Interdielectric applications have been considered as well. Silicone-containing polyimides exhibit low stresses in structures with silicon and metals because of their low moduli. However, they have high CTEs (coefficients of thermal expansion). Outgassing and solvent sensitivity are of concern as well.

Another block copolymer, polyimide-polyphenylenequinoxaline (PPQ)[38] block copolymers have been developed to improve the adhesion characteristics of polyimides. For various reasons these materials have not found application in the industry.

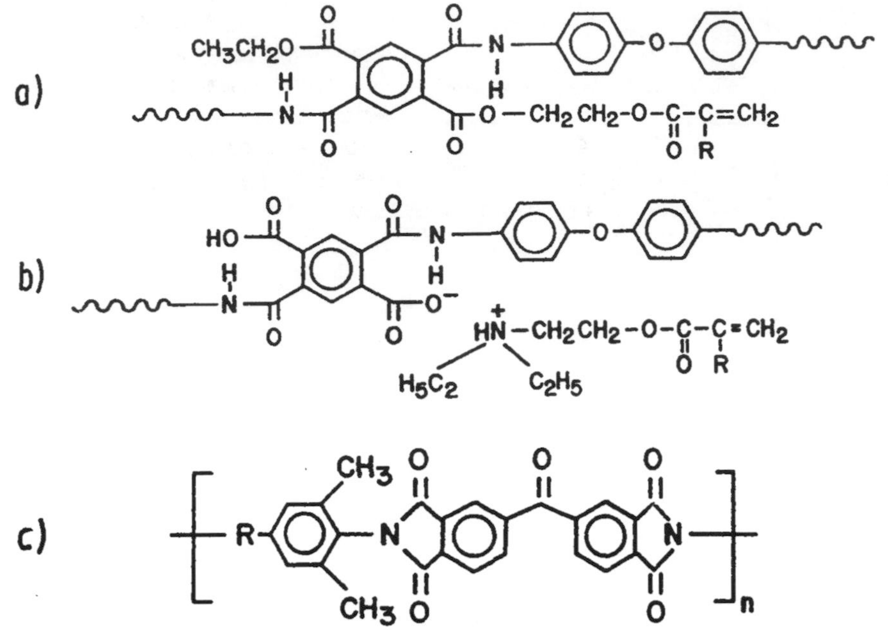

a)

b)

c)

FIGURE 12
Photosensitive polyamic acid ester (a), polyamic acid salt (b), and inherently photosensitive polyimide (c).

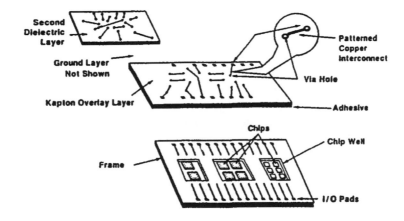

FIGURE 13
The GE process. (From Feger, C. and Franke, H., Polyimides in High-Performance Electronics Packaging and Optoelectronic Applications, in *Polyimides: Fundamentals and Applications*, Gosh, M. K. and Mittal, K. L., Eds., Marcel Dekker, New York, 1996, 759. With permission.)

A very advanced concept to lower the dielectric properties of inter-layer dielectric materials involves yet another polyimide block copolymer, polyimide (PI)-poly(propylene oxide) block copolymer.[39] These multi-block copolymers have very small (100 Å) domains of poly(propylene oxide) (PPO), uniformly distributed in the polyimide matrix. At elevated temperatures the PPO block unzips leaving uniform, close-celled pores[40] (Figure 14). The aim of this ongoing development is to obtain high temperature stable, mechanically strong interlayer dielectrics with dielectric constants significantly below 2.5.

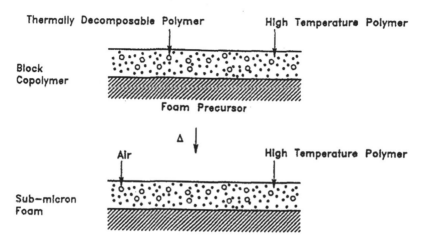

FIGURE 14
Closed cell foam formation through thermal decomposition of poly(propylene oxide) blocks. (From Hedrick, J., IBM Almaden Research Center, personal communication. With permission.)

One of the earliest polymers used by Japanese computer manufacturers for interlayer dielectrics has been polyisoindoloquinoline (PIQ),[41] which has the structure given below:

It cures in two steps, first forming the cycloimide ring and then the indolone ring. Compared to today's leading polymer dielectric, PIQ has a relatively high CTE.

The most recent, pure polyimides use an increased fluorine content to lower the dielectric constant while incorporating rigid-rodlike units to decrease the CTE. Two examples, PMDA-TFMB and 6FCDA-

TFMB,[42,43] where TFMB = 3,3'-bis(trifluoromethyl) benzidine, are given by the following:

(6FCDA-TFMB)

These polyimides exhibit relatively low dielectric constants and low CTEs. However, they tend to be anisotropic and expensive.

Another material that has excellent properties for interlayer dielectric applications is polyphenylenequinoxaline. Its main disadvantage is that it is only spin processible from *m*-cresol solutions and its dielectric constant is relatively high for today's demanding applications.

In the last few years thermosets intended for the interlayer dielectric market have been introduced. Most prominent are the benzocyclobutenes (BCBs) (Figure 15).[44-46] They exhibit excellent planarization, low dielectric constants, and excellent adhesion. However, the material is relatively oxidation sensitive which has been addressed by addition of an antioxidant package. Typical for thermosets these materials are relatively brittle but have been used successfully in the fabrication of MCMs.

A more recent development and a significant improvement over BCBs are the similarly named perfluorocyclobutanes (PFCBs).[47,48] The only other similarity with BCB is that they are thermosets as well. The chemistry is quite different (Figure 16). These materials have even lower dielectric constants (2.35) and also excellent planarization and adhesion. Again they are brittle compared to polyimides.

4. *Integrated Circuit Interconnects*

Figure 17 shows a multilayer metallization (MLM) as it is used for chip integration.[49] The analogy with a thin film module (MCM-D) is apparent. However, the dimensions of film and conductors are smaller by a factor of 50 to 100 ×. Current MLMs use three layers with SiO_2 as dielectric and aluminum as conductor metal. Increasing the number of interconnect levels will increase the speed of a central processing unit (CPU). Decreasing the dielectric constant of the dielectric and the electrical resistance of the conductor will increase the speed of the CPU further (Figure 18).[50]

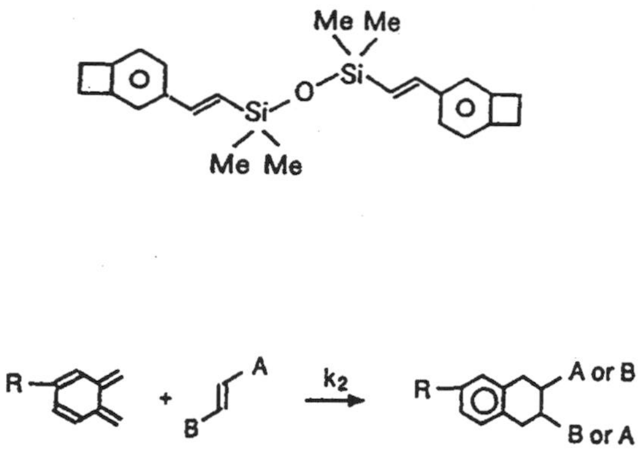

FIGURE 15
Chemical structure of a BCB monomer and one of the possible reactions of BCBs.

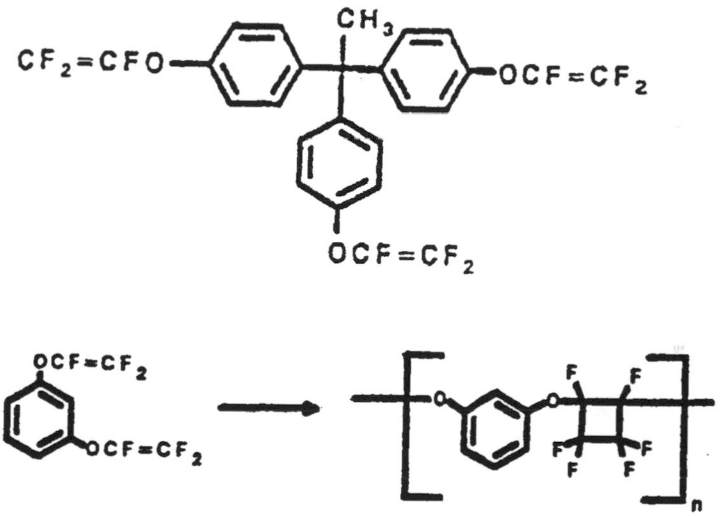

FIGURE 16
Chemical structure of trifunctional monomer and reaction of PFCBs.

The materials considered for MLM applications are essentially the same as those which have been considered for electronic packaging applications, particularly for MCM applications. An additional choice is Teflon AF, an amorphous fluorocarbon with very low dielectric constant (1.89). Issues that need to be resolved for successful application include adhesion of Teflon AF to deposited metal.

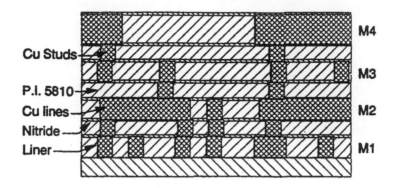

FIGURE 17
Multilayer metallization (MLM) interconnect structure. (From Harper, J. M. E., Colgan, E. G., Hu, C. K., Hummel, J., Buchwalter, L. P., and Uzoh, E. C., *Mater. Res. Soc. Bull.*, 19, no. 8, 23, 1994. With permission.)

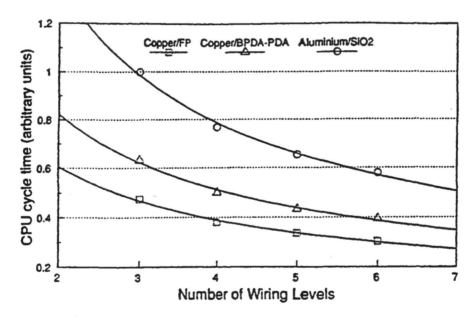

FIGURE 18
Relative CPU speed using a variety of metal/dielectrics combinations. (From Paraszczak, J., Edelstein, D., Cohen, S., Babich, E., and Hummel, J., *Proc. IEEE Int. Electron Devices Meet.*, 39, 261, 1993. With permission.)

IV. OPTOELECTRONIC APPLICATIONS

New developments in information transmission such as the information highway and video-on-demand have put new emphasis on the interface between electrically disseminated and optically transported information. This interface in which electrical information is transformed to optical and vice versa is the arena of optoelectronics. Optoelectronics components may or may not be part of high-performance electronics packages. If they are part of high-performance packages, they need to withstand high temperatures particularly if C4s are being used as chip join technology. An example of a high-performance application where optical and electrical information interface is shown in Figure 19. The optical signal is brought directly under the flip chip where it is converted into an electrical signal.[51] Such applications might become important for clocking of a large system.

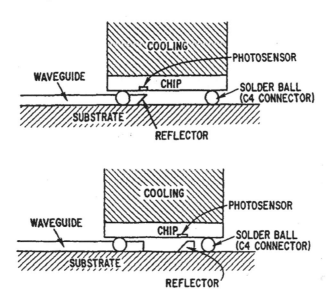

FIGURE 19
Two possible optical interconnects to a flip chip.

Materials that can be used to fabricate passive waveguides which carry the optical information to the chip are polyimides, photosensitive polyimides, Teflon AF, and PFCB among others. To be useful for such applications polymers need to exhibit optical thermal stability. This means the optical properties of the material must not change during exposure to high temperatures. This thermal stability is often not identical with the one measured by weight loss. For instance, the polyimide made from 6FDA-6FDAm shows a weight loss of less than 0.1%/h at 400°C. However, it turns brown during exposure to 350°C under nitrogen. At 300°C its optical properties do not change over many hours.[52]

V. ELECTRICALLY CONDUCTING POLYMERS

Electrically conducting polymers hold interesting promise for the electronics industry. A possible application would be use as anisotropically conductive adhesives. Such materials could allow low temperature (<150°C) bonding which would lower cost and could lead to improved mechanical connections. Before successful application, however, obstacles such as low conductivity compared to current Pb solder, z-axis movement during exposure to temperature and humidity, and questions of reworkability and durability need to be overcome. Polymer thick films (PTF) can be screen printed to form conductive, resistive, and dielectric elements on printed circuit boards and flexible circuitry.[53-55] Many of these polymer systems can be soldered to directly. Advantages include low cost and mechanical durability. Electronic applications include replacing wires on PCBs as an inexpensive, robust, and aesthetically pleasing means for engineering changes; and resistor-to-ground terminators for plug pin sockets.

VI. CONCLUSIONS

The significance and number of key applications of high temperature polymers in computer electronics will continue to grow. However, increased demand will be placed on cost and performance, requiring cooperative research through industry–government–academic consortia. In semiconductors and electronic packaging, advances in low dielectric, high temperature stable polymeric materials and photoresists will have a major impact thus furthering the trend of importance of polymers for high-tech areas such as in batteries, portable computers, and displays.

ACKNOWLEDGMENTS

The authors would like to thank Bob Allen, Maggie Best, Jeff Hedrick, Gerard Kopcsay, Jeff Labadie, Kostas Papathomas, and Jane Shaw, all of IBM, for sharing their work and for their input in clarifying some of the issues presented in this chapter. Appreciation is expressed to Barrie Jones of PolyMore Circuit Technologies for sharing his experiences with polymer thick films (PTF).

REFERENCES

1. Reisch, M. S., Chem. Eng. News, 24, August 30, 1994.
2. Freitag, D., in Prog. Polym. Sci., 19, 995, 1994.
3. Wirth, J. G., in Prog. Polym. Sci., 19, 947, 1994.

4. Marchant, A. B., Optical mass data storage II, *SPIE Proc.*, 695, 270, 1986.
5. Siebourg, W., Schmid, H., Rateike, F. M., Anders, S., and Grigo, U., *Polym. Eng. Sci.*, 30, 1133, 1990.
6. Tummala, R. R. and Rymaszewski, E. J., Eds., *Microelectronics Packaging Handbook*, Van Nostrand Reinhold, New York, 1989.
7. Papathomas, K. I., Japp, R. M., and Kohut, G. P., *Polym. Prepr.*, 36, no. 2, 366, 1995.
8. Hedrick, J. C., Gotro, G. T., Viehbeck, A., Proc. 7th Int. SAMPE Electronics Conf., June 1994, 171.
9. Rymaszewski, E. J., Walsh, J. L., and Leehan, G. W., *IBM J. Res. Dev.*, 25, 603, 1981.
10. Shambrook, K. P., Proc. of IEEE-CHMT Multichip Module Technologies and Alternatives Semin., San Diego, CA, April 13, 1991.
11. Doane, D. A. and Franzon, P. D., Eds., *Multichip Module Technologies and Alternatives: The Basics*, Van Nostrand Reinhold, New York, 1993.
12. Mukai, K. and Saiki, A., *IEEE J. Solid State Circuits*, SC-13, 462, 1978.
13. Ho, C. W., Chance, D. A., Bajorek, C. H., and Acosta, R. E., *IBM J. Res. Dev.*, 26, 286, 1982.
14. Koopman, N., Reiley, T., and Totta, P., *ISHM*, October 1988.
15. Rymaszewski, E. J. and Tummala, R. R., in *Microelectronics Packaging Handbook*, Tummala, R. R. and Rymaszewski, E. J., Eds., Van Nostrand Reinhold, New York, 1989, 1.
16. Jensen, R. J., in *Multichip Module Technologies and Alternatives: The Basics*, Doane, D. A. and Franzon, P. D., Eds., Van Nostrand Reinhold, New York, 1993, 225.
17. Feger, C. and Feger, C., in *Multichip Module Technologies and Alternatives: The Basics*, Doane, D. A. and Franzon, P. D., Eds., Van Nostrand Reinhold, New York, 1993, 311.
18. Feger, C. and Franke, H., in *Polyimides: Fundamental Aspects and Technological Applications*, Ghosh, M. K. and Mittal, K. L., Eds., in preparation.
19. Wilson, D., Stenzenberger, H. D., and Hergenrother, P., Eds., *Polyimides*, Chapman & Hall, New York, 1990.
20. Bessonov, M. I., Koton, M. M., Kudryavtsev, V. V., and Laius, L. A., *Polyimides, Thermally Stable Polymers*, Consultants Bureau, New York, 1987.
21. Boese, D., Herminghaus, S., Yoon, D. Y., Swalen, J. D., and Rabolt, J. F., *Mater. Res. Soc. Symp. Proc.*, 227, 379, 1991.
22. Edelstein, D. C., Scheuerman, M., and Geppert, L., Proc. 10th Int. VLSI Multilevel Interconn. Conf., Santa Clara, CA, June 1993, 511.
23. Ree, M., Chen, K.-J., Kirby, D. P., Katzenellenbogen, N., and Grischkowsky, D., *J. Appl. Phys.*, 72, 2014, 1992.
24. Feger, C., unpublished results.
25. Volksen, W., Yoon, D. Y., Hedrick, J. L., and Hofer, D., *Mater. Res. Soc. Symp. Proc.*, 227, 23, 1991.
26. Landis, A. L., *Polym. Prepr.*, 15, 537, 1984.
27. Rubner, R., Bartel, W., and Bald, G., *Siemens Forsch. Entwickl.*, 5, 235, 1976.
28. Angelopoulos, M., Gelorme, J., Swanson, S., and Labadie, J., *Soc. Plast. Eng. Technol. Pap.*, 40, 2165, 1994.
29. Asano, M., Eguchi, M., Kusano, K., and Niwa, K., Proc. IEEE/CHMT Eur. Int. Electr. Manuf. Technol. Symp., 213, 1993.
30. Pfeifer, J. and Rohde, O., in Recent Advances in Polyimide Science and Technology, Weber, W. D. and Gupta, M. R., Eds., Mid-Hudson Sect. of Soc. Plast. Eng., Poughkeepsie, NY, 1987, 336.
31. Maw, T. and Hopla, R. E., *Mater. Res. Soc. Symp. Proc.*, 227, 205, 1991.
32. Kowalczyk, T. C., Kosc, T., Singer, K. D., Cahill, P. A., Saeger, C. H., Meinhardt, M. B., Buehler, A. J., and Wargowski, D. A., *J. Appl. Phys.*, 76, 2505, 1994.
33. O'Brien, K., *Surface Mount Technol.*, 7, no. 5, 32, 1993.
34. Gdula, M., Wells, K. B., II, and Wojnarowski, R. J., *Digital Signal Proc.*, 2, 247, 1992.
35. Narayan, C., Purushothaman, S., Doany, F., and Deutsch, A., *IEEE Trans. Comp. Pack. Manuf. Technol. B*, 18, 42, 1995.

36. Davis, G. C., Heath, B. A., and Gildenblat, G., in *Polyimides: Synthesis, Characterization, and Applications*, Mittal, K. L., Ed., Plenum Press, New York, 1984, 847.
37. Saraf, R. F., Feger, C., and Cohen, Y. C., in *Advances in Polyimide Science and Technology*, Feger, C., Khojasteh, M., and Htoo, M., Eds., Technomics, Lancaster, PA, 1993, 433.
38. Hedrick, J. L., Labadie, J., and Russell, T., *Macromolecules*, 24, 4559, 1991.
39. Hedrick, J. L., Labadie, J., and Russell, T., Wakharkar, V., and Hofer, D., in *Advances in Polyimide Science and Technology*, Feger, C., Khojasteh, M., and Htoo, M., Eds., Technomics, Lancaster, PA, 1993, 184.
40. Hedrick, J., IBM Almaden Research Center, personal communication.
41. Satou, H., Suzuki, H., and Makino, D., in *Polyimides*, Wilson, D., Stenzenberger, H. D., and Hergenrother, P., Eds., Chapman & Hall, New York, 1990, 227.
42. Cheng, S. Z. D., Arnold, F. E., Jr., Zhang, A., Hsu, S. L., and Harris, F. W., *Macromolecules*, 24, 5856, 1991.
43. Auman, B. C., in *Advances in Polyimide Science and Technology*, Feger, C., Khojasteh, M., and Htoo, M., Eds., Technomics, Lancaster, PA, 1993, 15.
44. Kirchhoff, R. A., Carriere, C. J., Bruza, K. J., Rondan, N. G., and Sammler, R. L., *J. Macromol. Sci. Chem.*, A28, 1079, 1991.
45. Garrou, P. E., Heistand, R. H., Dibbs, M. G., Mainal, T. A., Mohler, C. E., Stokich, T. M., Townsend, P. H., Adema, G. M., Berry, M. J., and Turlik, I., *IEEE Trans. Comp. Hybr. Manuf. Technol.*, 16, 46, 1993.
46. Laursen, K., Hertling, D., Berry, N., Bidstrup, S. A., Kohl, P., and Arroz, G., *IEPS*, 2 (September 12–15), 751, 1993.
47. Babb, D. A., Ezzell, B. R., Clement, K. S., Richey, W. F., and Kennedy, A. P., *J. Polym. Sci. Polym. Chem. Ed.*, 31, 3465, 1993.
48. Perettie, D. J., Bratton, L., Bremmer, J., and Babb, D., *SPIE Proc.*, 1911, 15, 1993.
49. Harper, J. M. E., Colgan, E. G., Hu, C. K., Hummel, J., Buchwalter, L. P., and Uzoh, E. C., *Mater. Res. Soc. Bull.*, 19, no. 8, 23, 1994.
50. Paraszczak, J., Edelstein, D., Cohen, S., Babich, E., and Hummel, J., *Proc. IEEE Int. Electr. Devices Meet.*, 39, 261, 1993.
51. Feger, C., Perutz, S., Reuter, R., McGrath, J. E., Osterfeld, M., and Franke, H., in *Polymeric Materials for Microelectronic Applications*, Ito, H., Tagawa, S., and Horie, K., Eds. ACS Symp. Ser., 579, 1994, 272.
52. Reuter, R., Franke, H., and Feger, C., *Appl. Opt.*, 27, 4565, 1988.
53. Takeda, K., Umatsu, Y., Ohmasa, M., and Katoh, K., Proc. Jpn. Int. Electron. Manuf. Technol. Symp., Tokyo, June 26–28, 1991.
54. Prime, R. B., *Polym. Eng. Sci.*, 32, 1286, 1992.
55. Jones, B., PolyMore Circuit Technologies, 1995.

F

Chapter **8**

HIGH TEMPERATURE FIBERS: PROPERTIES AND APPLICATIONS

R. S. Irwin

CONTENTS

I. Introduction 149

II. Structure and Composition 151

III. Aramids 153

IV. Applications 154
    A. Flame-Resistant Clothing Systems 155
    B. Hot Gas Filtration 156
    C. Paper Applications 156
    D. Protective Clothing 157
    E. Miscellaneous 157

V. Melt-Processible High Temperature Fibers 158

VI. Conclusions 160

I. INTRODUCTION

High temperature fibers provide useful mechanical property levels at elevated temperatures or after prolonged exposure at high temperatures, e.g., 200°C, in air. They are characterized by high melting or softening

0-8493-7672-6/97/$0.00+$.50
© 1997 by CRC Press, Inc.

temperatures, and are dimensionally and chemically stable at high use temperatures. The structural features favoring high-temperature stability frequently coincide with those for flame resistance or high tensile strength and modulus. Table 1 contrasts selected thermal characteristics of m- and p-aramids, as typical high-temperature fibers, with conventional 66 nylon and poly(ethylene terephthalate) (PET) fibers and the intermediate Teflon® TFE fluorocarbon fibers. Typical of the aramid fibers are poly(m-phenylene-isophthalamide) and poly(p-phenylene-terephthalamide), sold as, for example, Nomex® and Kevlar®, respectively, by DuPont. Flammability may be conveniently compared in terms of limiting oxygen index (LOI) which is the percentage of oxygen needed in an oxygen/nitrogen atmosphere to just sustain vertical burning. Since air contains 21% oxygen, materials with LOI above 21 are not readily flammable in air.

TABLE 1.

Comparison of High-Temperature and Conventional Fibers

	% Strength Retained at 260°C	Time for 50% Strength Loss at 180°C in Air	Melting Point	Shrinkage at 180°C in Air	Flammability LOI
66 Nylon	Melts	100 h	254°C	4–11%	24 (Melts)
PET	Melts	300	256	3–12	23 (Melts)
Teflon® TFE	6%	Years	310	7	95
m-Aramid	58	Years	410	0.4	29
p-Aramid	63	Years	None	<0.1	29

Worldwide manufacturing capacity for high-temperature fibers, including high-strength varieties is probably close to 100 million lb/year, representing a value of more than $1.0 billion. The average selling price of $10 to $15/lb is high relative to most other fibers. m- and p-Aramids account for probably 75% of total capacity, with the latter even more dominant in the high-strength fiber area. Aramids were identified in the 1950s as representing an optimal balance of properties, processibility, and economics. By installing sizable capacity at an early stage, DuPont, in particular, was ultimately able to benefit from economies of scale in both intermediates and fiber production to assure a dominant market position today. Since aramids, as high-temperature or high-strength fibers, initially were not replacing incumbent materials, a great deal of applications research was necessary to seek out and adapt aramid fibers for optimal effect in new uses. No particular application today is heavily predominant; rather, high-temperature fibers have found their way into a myriad of small-volume applications, some quite unusual and with remarkable high visibility, despite small size.

Beyond aramids the high-temperature fiber market is shared by a variety of other compositions, manufactured at relatively small capacities of little more than a few million annual pounds. These frequently find their way into niche markets emphasizing one or another special property. For example, polybenzimidazoles and polypyromellitimides provide enhanced high-temperature stability relative to aramids, while polyphenylene sulfide has enhanced chemical inertness. Blending with other high-temperature fibers is often used to achieve balances with other desired properties. High-temperature fiber products are currently manufactured in the form of continuous multifilament yarns, crimped staple, floc (very short fibers), fibrids (tiny filmlike particles), papers, and nonwovens.

II. STRUCTURE AND COMPOSITION

Apart from flame protection applications, high-temperature fibers must maintain mechanical and chemical integrity up to high temperatures. The usual chemical changes leading to cross-linking and chain scission are commonly manifested by progressive embrittlement and ultimately weight loss. For maximum thermal stability, polymers containing a high frequency of stable, aromatic, or heteroaromatic rings, with minimization or elimination of aliphatic components, are indicated.

The key to high-temperature dimensional stability is to combine a high glass transition temperature (T_g) (which often heralds a sharp change in mechanical properties) with substantial crystallizability. In cases of very high T_g it becomes less important to have a high degree of crystallinity. Since T_g is usually 67 to 80% of the crystalline melting point (T_m in Kelvin), it is implicit that T_m should be as high as possible. T_m is related to the heat of melting (ΔH) and entropy of fusion (ΔS):

$$T_m = \frac{\Delta H}{\Delta S}$$

T_m may be increased by increasing ΔH, i.e., intermolecular forces, by such means as hydrogen bonding, polar groups (heteroatoms) and by lowering ΔS, i.e., by increasing chain stiffness. The latter would involve increased frequency and size of aromatic or heteroaromatic ring units, emphasizing para-orientation, and short bridging groups such as ether, sulfide, carbonyl, and more especially the linear ester or ether groups. An exception to emphasis on aromatic structures for high chemical stability is poly(tetrafluoroethylene) and related fluorocarbons, but these tend to soften and lose strength at high temperatures.

A selection of commercial high-temperature fiber structures, illustrating the foregoing principles, is provided in Table 2. These, perhaps, illustrate the trend away from these compositions which provide the very best high-temperature performance but are expensive, toward less expensive, adequate compositions. Polypyromellitimides and polybenzimidazoles represent the former, while the recently introduced BASF Basofil® illustrates the latter. Basofil®, from inexpensive melamine and formaldehyde, is directed toward providing flame protection at minimum cost, without having the wide range of applicability and other properties afforded by

TABLE 2.

Commercial High Temperature Fiber Compositions

Type	Structure	Trade name (Company)	Temp.
Meta-aramid	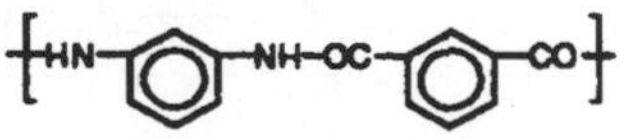 [HN—⬡—NH—OC—⬡—CO]	Nomex® (DuPont) Teijinconex® (Teijin)	220°C
Para-aramid	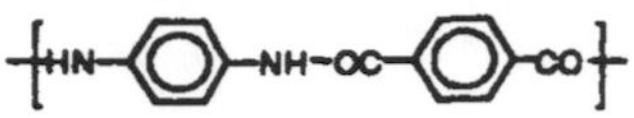 [HN—⬡—NH—OC—⬡—CO]	Kevlar® (DuPont) Twaron® (Akzo)	220°C
Polyamide-imide	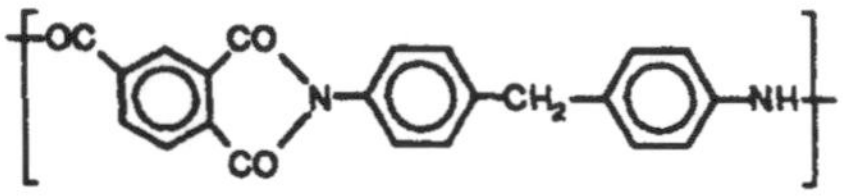 [OC/CO N—⬡—CH₂—⬡—NH]	Kernel (Rhone-Poulenc)	250°C (est.)
Polyimide	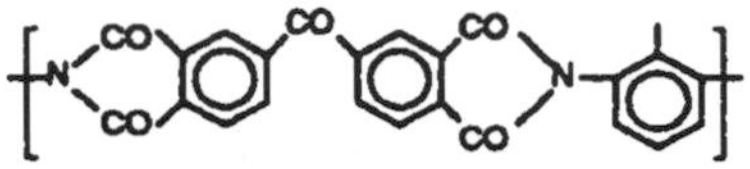 [N(CO)₂—⬡—O—⬡(CO)₂N—⬡] Copolymer	P84 (Lenzing)	250°C (est.)
	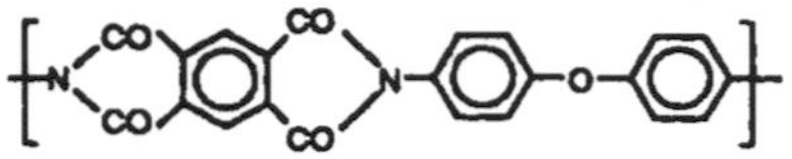 [N(CO)₂—⬡—N—⬡—O—⬡]	Aramid PM (Russia)	300°C (est.)
Polybenz-imidazole	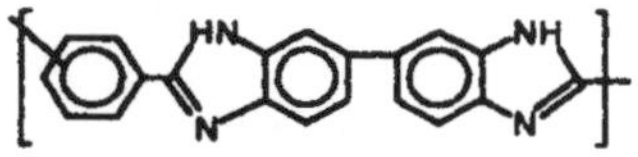 [⬡(HN)(N)—⬡—(NH)(N)]	PBI (Hoechst-Celanese)	>300°C
Novoloid	Phenol-Formaldehyde	Kynol (Japan)	200°C
	Melamine-formaldehyde	Basofil (BASF)	200°C (est.)
Polyphenylene Sulfide	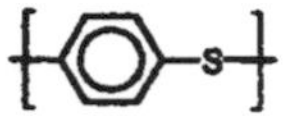 [⬡—S]	Ryton (Amoco Fibers)	175°C
Polytetrafluoro Ethylene	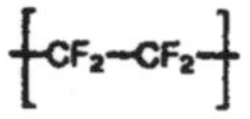 [CF₂—CF₂]	Teflon TFE (DuPont)	240°C

m-aramids. Very high-performance high-temperature polymer compositions, illustrated below, have hardly advanced beyond the research stage.

Few organic fibers are capable of prolonged service at 300°C or brief service at 400°C. Most promising were the ladder polymers consisting solely of fused aromatic or heterocyclic rings. Gains in high-temperature performance were almost invariably accompanied by increasing difficulty in forming the polymer into fibers, and increasing costs.

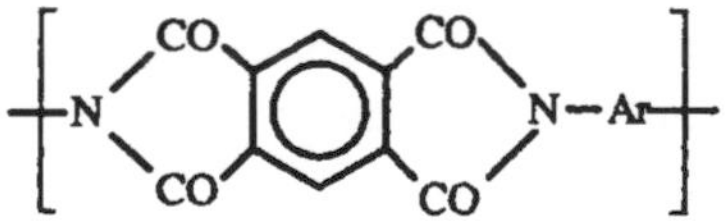

[N(CO)₂—⬡—(CO)₂N—Ar]   Polypyromellitimide

Polyoxadiazole

"PBX"
X = NH Polybenzimidazole
X = O Polybenzoxazoles
X = S Polybenzthiazoles

"Ladder" Polymer
(50% strength reduction
at 500°C)

With the end of the cold war, continuation of a high level of military and aerospace spending, including research on high-performance materials, has become subject to close scrutiny. At the same time, civilian interest in high-performance fibers has been limited, and tempered by high costs of materials. Additional to disincentives any new fiber candidate must face are the substantial expenditures required for testing and qualification, especially in the aerospace sector. Without a major change in properties or costs there exists considerable inertia to change from the tried and true, usually aramid systems.

III. ARAMIDS

An appreciation of the properties and applications of high temperature fibers in general may be best gained by a review of aramid technology, which represents the major focus of today's high-temperature fiber market.

Nomex® and other m-aramids are characterized by relatively flexible polymer chains which are spun into fibers by conventional solution-spinning methods. Fiber strength is built up by drawing the fibers at elevated temperatures to orient them in the axial direction and crystallize them. Kevlar® and analogous p-aramids are likewise synthesized by reaction of diacid chlorides with diamines in an organic solvent. In the case of p-aramid homopolymers, the polymer chains are extended in solution with the spontaneous formation of a liquid crystalline phase, so that by flow through a spinneret the molecular chains become highly aligned in the axial direction; and this arrangement is captured when the fibers are coagulated. Such a structure without further modification is the basis for very high-tensile strength and modulus, since applied loads are shared equitably among a high proportion of the molecular chains. The identical result may be achieved with certain aramid copolymers such as Technora® (Teijin), which are too flexible in solution to form a liquid crystalline phase, but wherein the as-spun fibers may be highly drawn to provide the desired structure.

Table 3 compares key properties of Nomex® and Kevlar® as typical *m*- and *p*-aramid fibers. The low elongation of *p*-aramids is a measure of limited toughness and is associated with fibrillar structure, resulting in limited resistance to flexing or abrasion.

TABLE 3.

Fiber Properties Representative of *m*- and *p*-Aramids

	m-Aramid	*p*-Aramid	Comment
Color	White	Yellow	Conjugation effect
T_g	265°C	ca. 425°C	
Crystallinity	Moderate	Very high	Low for copolymers
T_m	415°C (dec)	Nil up to 550°C	
Pyrolytic weight loss	410°C	500°C	In air
Continuous use ceiling	220°C	220°C	Embrittlement
Tensile strength	5 gpd	23–27 gpd[a]	
Initial modulus	90 gpd	500–900 gpd	Depends on crystallinity
Elongation at break	30%	3–5%	
Strength retention, 200°C/300°C	75/38%	75/50%	Short-term exposure
Shrinkage, water/100°C	2.0%	Negligible	
Shrinkage, air/300°C	4.0%	Negligible	
Moisture regain	4–6%	4–6%	65% Relative humidity
Fiber structure	Nonfibrillar	Fibrillar	
Compressive strength	—	< Carbon or glass	
Organosolubility	Very limited	Insoluble	

[a] grams per denier: c.f. nylon, ca. 8 gpd

In addition to poor wear resistance, the yellow color of *p*-aramids is not conducive to facile colorability or attractive appearance, while high stiffness of the fibers is antithetical to comfortable apparel. *p*-Aramids are much less suited to apparel applications than are *m*-aramids.

While *m*- and (especially) *p*-aramids show good thermal stability at quite high temperatures, the ceiling for long-term exposure is relatively modest, 220°C. Progressive embrittlement is attributed to a rearrangement reaction between amide groups in adjacent chains where one chain is split and the other becomes branched.

IV. APPLICATIONS

A broad survey of various applications of aramid fibers will provide an overview of the multifarious uses of high-temperature fibers. As mentioned, outside of a few larger applications such as flame protection, the market contains a great many small-volume applications. In developing many of these markets it has been necessary to integrate a variety of base technologies (e.g., polymer structure, fiber science, engineering applications development) in partnership with a customer. Even in cases where a new application is not being created but a material such as asbestos in gaskets or nylon in ropes or tire cord is being replaced, a thorough understanding is necessary to achieve maximum impact with the newer fiber.

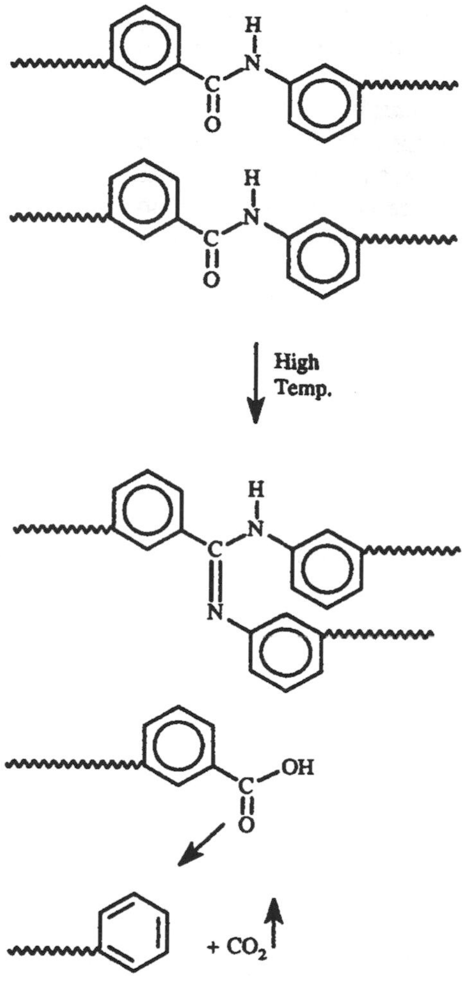

A. Flame-Resistant Clothing Systems

In a flash fire or electric arc, *m*-aramid clothing forms an expanded (50% larger volume) insulating charlike material which does not burn or melt-drip, thus minimizing burn injury, and providing insulation and escape time. The amidine rearrangement, shown above, accelerates greatly when *m*-aramids are raised quickly to the softening or melting point, and decarboxylation produces significant foaming; hence expanded volume. Beyond this point the material continues to char. Its brittleness and propensity to break open, is compensated by blending with *p*-aramid which enables the charred garment to retain its integrity against shrinkage or brittle disintegration. In an alternative approach the considerably more stable but more expensive polybenzimidazole, in place of *m*-aramid, retains flexibility and integrity to a greater extent. Enhanced protection is provided by use of multilayers of *m*-aramid clothing, i.e., under- as well as outerwear.

Acceptance in workplace protection clothing, military uniforms, firefighter clothing, flight suits, and the like requires comfort, good appearance, and durability as well as flame protection for the life of the garments. *m*-Aramid hydrophilicity, thus, is key to comfort, while color, colorability, crease resistance, chemical resistance, and antistatic character (in Nomex® IIIA, yarns contain a carbon core) all play an important role in fabric aesthetics.

Fiber blending is widely applied for adapting alternative flame-retardant fibers, e.g., polyamide-imide, polyimide, melamine-formaldehyde, or phenol formaldehyde to flame-protective clothing uses.

B. Hot Gas Filtration

To prevent ejection of solid particulate into the atmosphere, hot effluent gases from industrial operations are passed through filters which may consist, typically, of a woven or nonwoven felted fabric which may be attached to a strengthening scrim. (Nonwovens may be needle-punched or spun-laced by water jets.) As well as high-temperature resistance and gas permeability, these filter bags must provide attractive economics through frequent reusability. A common shortcoming is embrittlement caused by attack by aqueous sulfuric acid droplets present in certain off-gases.

C. Paper Applications

m-Aramid papers and pressboard consist of short fibers bound together by fibrids, which are small, fiberlike particles made by precipitation of polymer solutions in water. Papers are used for insulation in motors, generators, and other electrical equipment. *m*-Aramids may be kept in continuous use up to 220°C. Their toughness facilitates fabrication and guards against unexpected stresses. Thermount® contains *p*-aramid fibers with a *m*-aramid binder and is used for high-density circuit boards where great uniformity and smooth surface are provided, together with a very low coefficient of expansion (or contraction).

Three-dimensional honeycomb structures made from aramid papers are sliced into wafers of various thickness, and faced with adherent flat sheets to provide extremely lightweight but nevertheless rigid panels which find use, e.g., in aircraft interiors as paneling, flooring, and overhead bins.

D. Protective Clothing

p-Aramid fabrics have a remarkable resistance to cutting through — as much as tenfold that of leather for similar weight materials. Thus Kevlar® chaps are capable of stopping a chain saw at 3000 ft/min. This is associated with very high molecular alignment, such that a high density of covalent bonds must actually be parted for a knife to penetrate. Gloves, aprons, and limb guards thus find widespread application in the glass and steel industries, as well as elsewhere.

Molded p-aramid fabrics impregnated with toughened phenolic resins provide helmets for protection against ballistic threats such as shrapnel and bullets. These afford a weight advantage of 50 and 60%, respectively, vs. steel and nylon. Current materials are subject to improvement as tougher aramid yarns emerge. Competition is currently provided by ultrahigh molecular weight polyethylene high-strength yarns, e.g., Spectra® (Allied-Signal), which is quite low melting.

Bulletproof vests consisting of several layers of p-aramid fabrics, affording light weight and comfort, are used by various military and police departments. In the latter case, about 1700 lives are said to have been saved so far.

In steel mill rod handling (at 1000°F), p-aramids have proved effective as replacement for asbestos. Heat protection may be enhanced by aluminizing. p-Aramids provide effective protection against molten metal splashes.

E. Miscellaneous

As asbestos replacement, the use of p-aramids, in the form of pulp, also extends to gaskets for high-temperature use. In clutch plates and brake linings, additional use is made of the ability of Kevlar® to absorb energy.

An obvious application for p-aramids is as lightweight cables or ropes, e.g., for mooring oil well drilling rigs, and support for fiber optics cable. Even in such apparently straightforward uses ropes must be designed to allow maximum tensile strength while minimizing any adverse effects of susceptibility to abrasion. Aramids are used in a variety of reinforcement applications, e.g., drive belts, rubber hoses, automobile tires, and plastics reinforcement (often in combination with the more compression-resistant carbon, where the predominant applied stress is tensile). Many minor uses range from wrapping of electrical wires for use at high temperatures, sailcloth, reinforced sporting goods such as canoes, thixotropic additives to fluids or ironing board covers.

V. MELT-PROCESSIBLE HIGH TEMPERATURE FIBERS

The drive toward minimum adequate properties at lowest cost for a given end use has diverted some attention to the proposition that melt-spun fibers might be adequate for certain purposes, e.g., filter bags operating well short of 200°C. The incentive here is minimum investment in equipment because solvent handling, containment from the environment, recovery, losses, and disposal are avoided. Melt processing at high temperatures presents several difficulties, none of which have been resolved at this time.

To reduce the melting point of polymeric compositions which are, by nature, high melting, it is necessary to make adjustments toward a less ordered structure by copolymerization, incorporation of flexible ether or *m*-phenylene groups, or the like. These changes often have the effect of reducing or inhibiting crystallizability with deleterious consequences for high-temperature dimensional stability and possibly other properties, as well as affecting the cost picture. Melt-spinnable polyetherketones may be

regarded as copolymers containing both isophthaloyl and terephthaloyl groups, which, while reducing melting point to 350°C or less, drops T_g to almost 160°C; and more seriously, reduces degree of crystallizability. Dimensional stability, therefore, falls off not far beyond T_g. Another example is Ultem® (General Electric), a polyimide-ether which, by virtue of its flexible, nonlinear chain structure, can be melt-spun, while retaining $T_g = 215$°C.

This material does not crystallize at all and therefore loses fiber properties unacceptably approaching the T_g.

Recently Mitsui–Toatsu introduced a melt-spinnable polypyromellit-imide (Aurum®) which does have a remarkable propensity to crystallize:

The melt-processing temperature at 388°C is much higher than that of conventional polymer melts, but nevertheless is below the decomposition threshold; T_g 250°C, is high, so that in crystalline form, high-temperature dimensional stability is good. Aurum® thus does not shrink in a flame, although it ultimately melts. However, Aurum® illustrates a second major problem. Aromatic polymer chains, especially with fused polycyclic components, are inherently a good deal stiffer than aliphatic chains like 66 nylon. In order to maintain melt viscosity at processible levels in such polymers, it is necessary to reduce molecular weight considerably. While this may not necessarily reduce fiber tenacity unacceptably, resistance to flexing abrasion sharply deteriorates. As examples, standard fibers such as 66 nylon or Nomex®, having a degree of polymerization on the order of at least 100 commensurate with good mechanical properties, may show a flex life of 60,000 cycles or more. Under the same conditions, a melt-spun polyether-imide, having a necessarily much reduced degree of polymerization, may show a flex life of under 5000 cycles, and thus calls into question its ability to provide an acceptable level of wear resistance.

In the case of thermotropic liquid crystalline polyesters, the unusually high axial alignment of the molecular chains allow the as-extruded fibers to be heated close to their melt temperatures without severe shrinkage as with conventional fibers. Under these conditions polymerization can continue in the solid state such that by subjecting as-spun fibers to a suitable heat treatment, preferred levels of molecular weight and mechanical performance can be achieved. Thus heat treated, these fibers provide properties similar to Kevlar®, and candidacy for similar applications where meltability is not a deterrent. However, T_g of 110 to 170°C is rather low but despite copolymer character and modest levels of crystallinity, these fibers are dimensionally rather stable. Thermotropic polyester fibers have been slow to gain acceptance in the face of the well-established p-aramids. Best known is Vectra® (Kuraray licensed from Hoechst–Celanese) which is a copolymer of p-hydroxybenzoic acid and 6-hydroxy-2-naphthoic acid. At this point manufacturing capacity is of the order of 1.0 million lb annually.

VI. CONCLUSIONS

The dominant incumbents in the high temperature field are *m*- and *p*-aramids. These were originally selected for manufacture on the basis of an optimum balance of properties, applicability, and cost. Relatively large manufacturing facilities were built years ago by DuPont for ingredients and fibers, to realize economics of scale. Over many years a multitude of applications were developed, with emphasis on entire systems. The relative importance of pulp uses has grown strongly. Today, worldwide, there are several aramids manufacturers.

Costly qualification for various applications and increased familiarity with aramid products have created inertia toward use of alternatives, e.g., in military areas. There are no clear-cut cost advantages for any particular alternative type of high temperature fibers over aramids, but there is a consciousness that for certain applications aramids are overengineered and that cheaper alternatives may be adequate for certain uses. Niche markets have been perceived for alternative compositions, but these are subject to cost disadvantages of limited scale.

Activity on very high-performance high temperature fibers has abated in view of their high cost, and lower demands for military and aerospace applications.

The high temperature industry may be viewed as relatively mature and expanding slowly, but this is subject to significant alteration, e.g., via cheaper routes to ingredients or increase in military demand.

Chapter 9

PBI FIBER AND POLYMER APPLICATIONS

George A. Serad

CONTENTS

Abstract 162

I.   Introduction 162

II. Personal Fire and Flame Protection.......................................163
     A.   Definition of the Hazard 163

III. Properties of PBI Fiber 166

IV. Properties of PBI-Containing Fabrics...................................166

V.   Performance of PBI Fabric Blends 168

VI. Emerging Applications and Technology 173
     A.   Electrochemical Applications of PBI............................174
          1.   Fuel Cells...174
          2.   Batteries..175
     B.   Polymer Blends 175
          1.   Plastic Parts 175
          2.   Polymer Alloy Fibers 176
     C.   Catalysis 176

VII. Conclusions 177

Acknowledgment..177

References...177

ABSTRACT

Polybenzimidazole (PBI) fiber and polymer have been commercially available since 1983. The unique high temperature, nonmelting, nonburning and high moisture regain properties have stimulated a strong focus on the use of PBI fiber in personal fire and flame protection. PBI fiber properties, fire hazards, and resultant applications are reviewed. Some potential applications for PBI in other than fiber form are also noted.

I. INTRODUCTION

In 1985 the Federal Trade Commission approved a new generic fiber classification designated as polybenzimidazole (PBI) to represent, "A manufactured fiber in which the fiber forming substance is a long chain aromatic polymer having recurring imidazole groups as an integral part of the polymer chain". The only commercial facility to manufacture PBI fiber is operated by Hoechst Celanese Corp. at its plant in Rock Hill, South Carolina which started operation in 1983.

Polybenzimidazoles are prepared by the condensation of tetraamino compounds with diacids. Although a variety of aliphatic and aromatic diamines have been explored, the preferred polymer is poly[2,2'-(m-phenylene)-5,5'-bibenzimidazole] which is prepared from the monomers 3,3',4,4'-tetraaminobiphenyl and diphenyl isophthalate. A detailed review of the preparation and properties of polybenzimidazoles is provided in References 1 and 2.

Important attributes of PBI fiber are that it does not burn in air and has excellent high temperature properties. For example, decomposition does not occur until the temperature exceeds 580°C in air and 700°C in nitrogen. In addition PBI retains a high proportion of its physical properties at very high temperatures. Solid-state deuteron nuclear magnetic resonance spectroscopy experiments used to investigate the mobility of specifically labeled isophthalate rings in PBI-d_4 at temperatures above 400°C showed little evidence of mobility.[3] Further, it does not melt, contribute fuel to flames, or produce smoke. When exposed to a flame it forms a tough residual char. In nonoxidizing environments, PBI will carbonize in yields up to 80% and retain its form and flexibility as it carbonizes. PBI also retains its physical properties at cryogenic temperatures, e.g., at liquid nitrogen temperatures, where its properties, including toughness, are similar to those at room temperature.

Although PBI resists most solvents, it can be dissolved under specific conditions with a few polar aprotic solvents. Although this permits the fabrication of various PBI polymer forms, the primary commercial focus of PBI is as a fiber. This is due to its unique balance of properties which provides an excellent personal protective flame barrier. Although this discussion will primarily focus on flame and fire protection using PBI fiber, a few additional PBI properties and applications will be highlighted.

II. PERSONAL FIRE AND FLAME PROTECTION

The need for personal flame and heat protection is indicated by the number and severity of burn injuries that occur each year. Over 100,000 individuals are hospitalized each year with burns. Half of them result in substantial temporary or permanent disability, and over 6000 people die from the injuries. Survival of these burn victims is dependent on the extent of the burns as well as their age and physical condition. For example, someone with 60% body burns has a 80% chance of survival in the 5 to 34 age group, 60% in the 35 to 49 age group, and 40% in the 50 to 60 age group. Of course the degree of permanent disability goes well beyond these figures.

Approximately 20 to 25% of the burn injuries occur in the workplace. In addition to the individual's trauma and the time lost from work, the cost of medical services for a severe burn victim requiring a period of intensive care treatment is over $250,000. The magnitude of this cost alone represents a significant justification for providing adequate safety protection.

A. Definition of the Hazard

To provide adequate personal protective apparel it is important to define the type of fire or flame hazard that might be encountered in the particular activity. Key parameters are the duration of the exposure and the heat flux. Three distinct groupings are burning clothing, flash fire, and arc flash (Figure 1). The burning clothing scenario is usually the first that comes to mind. In this case exposure to an open flame causes the clothing to start burning. Exposure periods of 30 to 90 s are common. The next level of severity would be a flash fire, i.e., a sudden eruption of flame of high intensity that would usually last for 1 to 6 s. Such an occurrence might result in a fire-fighting situation when conditions exist to ignite a spontaneous fireball. The third hazard, an arc flash, is of even shorter duration (e.g., tenths of a second) but of higher intensity. An example would be the occurrence of an electrical flash encountered with high-voltage power lines.

In the burning clothing hazard, the temperature would be above the ignition or melting temperature of common garment fibers, e.g., cotton, polyester/cotton, nylon, or wool (Figure 2). Where such a hazard is a possibility, the garment should be constructed from a fabric that does not burn or form molten polymer drops. However, it is important to recognize that a nonburning fabric alone is not sufficient. Many routinely used materials may pass flammability tests (e.g., FSTM-191-5903, FSTM-191-5905, thermal protective performance [TPP]) but do not provide continuing protection once exposed to flame. Many form a hard, crusty char that help them perform well on static burn tests, but easily fracture in a dynamic situation (particularly if the individual is moving to escape the fire) and expose the individual to the flames. PBI has a unique ability to

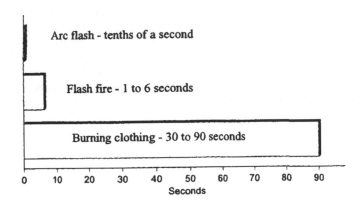

FIGURE 1
Comparison of duration time for fire and flame hazards.

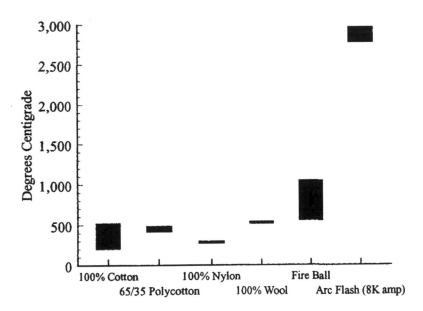

FIGURE 2
Approximate ignition and melting temperatures of various fabrics compared to the temperature of a fireball or arc flash.

form a tough, flexible char that continues to provide protection even when an individual is moving to escape the fire.

In the flash fire scenario the heat flux can be in the 1 to 10 cal/cm²/sec range. Exposure to 5 cal/cm²/sec for 1 s can cause third-degree burns. Here it is important that the protective garment both survive the exposure and provide continuing protection.

The arc flash situation is perhaps the most severe. A total heat flux of up to 30 cal/cm² for a few tenths of a second is possible. Portions of this energy must be dissipated to avoid severe burns. The potential severity is dependent on the proximity of the individual to the arc (Figure 3). As an example, an individual 12 in. away from an arc flash of 10 cal/cm² would need to have about 5 cal/cm² dissipated by the protective garment or surroundings to avoid a third-degree burn, another 3 cal/cm² to avoid a second-degree burn, and another 2 cal/cm² to avoid a first degree burn.

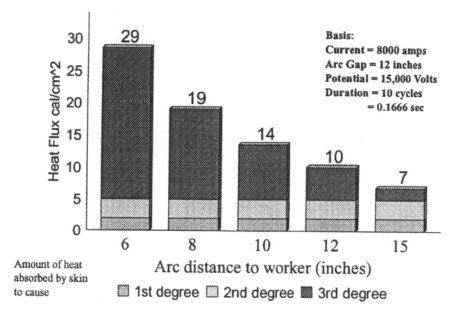

FIGURE 3
Heat flux generated by an arc flash and anticipated extent of burns.

Fire- and flame-resistant fabrics and garments are used throughout a variety of industries (Figure 4). An increase of 25% in the consumption of flame-resistant fabrics, from a 1992 base of 24.5 million yd²/year, is projected over the 1992–1997 time frame. The largest growth is projected for the utilities and refinery/chemical industries.

In the selection of personal protective apparel the first concern is to assure a base level of protection. This includes an assurance that the fabric will not readily burn, ignite, or melt drip; provides some thermal protection; and exhibits minimal flame shrinkage. Given that, there is a need to be certain that the protective gear is worn and does not interfere with the ability of the workers to perform their duties. This introduces the concept of comfort, which is actually a relative term since it is highly dependent on the environment and level of protection required. All of this is further influenced by cost. It is a frequent misconception to relate cost to lowest purchase price. In many cases, neither a life cycle approach nor the inclusion of potential liability and medical costs are included in the selection of personal protective apparel.

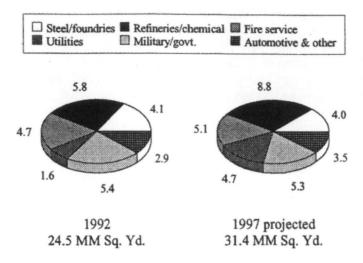

FIGURE 4

Industrial consumption of fire protective fabrics in the U.S.

III. PROPERTIES OF PBI FIBER

PBI is one of many commercial flame-resistant fibers and treated fabrics (Table 1). It is produced as a staple fiber that can be processed on cotton and worsted systems and can be blended with other fibers. Table 2 summarizes some of its physical and chemical properties. The commercial PBI fiber is actually a partially sulfonated polymer. This is done to reduce the fiber flame shrinkage so that it provides continuous protection even when exposed to flame. This key attribute also carries over into blends with other fibers so that their protective performance is enhanced.

Clearly, its inherent flame and thermal properties are impressive and superior to other fibers, but it is only the final fabric and protective garment that impacts the consumer. Here, the ability of PBI to transform its unique properties into 100% PBI fabrics and engineered blends with other fibers distinguishes its performance relative to other material alternatives.

IV. PROPERTIES OF PBI-CONTAINING FABRICS

Impressive fiber properties alone are not sufficient to produce a viable protective garment. It is necessary to engineer a cost-effective fabric and garment to meet the total needs of the consumer, e.g., protection, comfort, and cost. PBI textile fibers help produce effective fabrics and garments

TABLE 1.

Various Commercial Flame and Fire Protective Fibers and Fabrics

Manufacturer	Fiber/Fabric	Polymer	Comments
E.I. DuPont	Nomex	*m*-Aramid	
	Nomex III	*m*-Aramid	95/5% Nomex/Kevlar
	Nomex IIIA	*m*-Aramid	Nomex/Kevlar/Nega-stat
	Kevlar	*p*-Aramid	
Rhône-Poulenc (Amoco)	Kermel	Polyamideimide	50% Blend with FR Viscose
Teijin	Conex	*m*-Aramid	~Nomex
	Technora	*p*-Aramid	~Kevlar +
Akzo	Twaron	*p*-Aramid	~Kevlar
Hoechst Celanese	PBI	Polybenzimidazole	Blended with Kevlar/FR Rayon/Nomex
Westex	Vinex	PVA/polynosic	85/15% Blend
	Indura Proban	Fr Cotton-ammonia cure	Permanent FR
	Proban	Fr Cotton-ammonia cure	Less durable FR
Lenzing	P84	Polyimide	
	Viscose	Rayon	Phosphorous treated
	Wool	FR treated	Zirconium complex
UCC	Dynel	Modacrylic	Staple fiber
	Verel/Kanekalon /Teklan	Modacrylic	
Galey & Lord	Fabric — Flamex	FR blend of poly/cotton	
Springs Protective Fabrics	Fabric — Fire wear	55/45 Blend of cotton and a modified modacrylic	Acrylic modified with antimony oxide

TABLE 2.

Physical Properties of PBI Staple Fiber

Property	Value
Denier per filament	1.5 denier (1.7 dtex)
Tenacity	3.1 g/d (2.7 dN/tex)
Breaking Elongation	30%
Initial modulus	45 g/d (40 dN/tex)
Crimps per unit length	12/in. (0.5/mm)
Density	1.43 g/cm^3
Moisture regain @ 68°F and 65% RH	15%
Flame shrinkage	< 10%
Hot air shrinkage, 400°F	< 1%
Limiting oxygen index	38%

when used at levels from 20 to 100% depending on the specific application and the other fiber constituents. The PBI fiber impact on fabric properties is easily seen in several distinct areas:

- Improved mass and physical property retention during flame or thermal exposure yielding a dimensionally stable and flexible carbonaceous char
- Lighter weight fabrics because of the excellent thermal stability and fire protection characteristics of PBI
- Improved fabric flexibility and hydrophilicity leading to improved comfort
- Improved resistance to chemical and hydrolytic degradation

Capitalizing on these attributes, target markets for PBI-containing fibers are fire service, industrial and protective apparel, military apparel and special garments, fire blocking, and asbestos replacement. Special PBI blends have been developed for each of the specific markets to tailor the elements of protection, comfort, and cost. Some common PBI-containing commercial fabrics are shown in Table 3.

TABLE 3.

Principal PBI-Containing Fabrics Used for Fire and Flame Protection

Fabric Type	Construction	End Use
4.5 oz/yd² 40/60 PBI/p-aramid	2 × 1 Twill	Shirt/coverall
6.0 oz/yd² 40/60 PBI/p-aramid	2 × 1 Twill	Pant/coverall
6.0 oz/yd² 40/60 PBI/p-aramid	Rip stop	Turnout coat
6.0 oz/yd² 20/80 PBI/rayon	Knit	Sock hoods
6.0–7.5 oz/yd² 40/60 PBI/p-aramid	Nonwoven	Seat blocking
7.0 oz/yd² 40/60 PBI/p-aramid	Knit	Aluminized
7.5 oz/yd² 40/60 PBI/p-aramid	Rip stop	Turnout gear
10–24 oz/yd² 40/60 PBI/p-aramid	Various	Gloves/other

V. PERFORMANCE OF PBI FABRIC BLENDS

Several standard tests are used throughout the industry to characterize and qualify flame-resistant materials (Table 4). In most cases PBI is used in blends to obtain the best balance of cost performance. The performance of PBI blends relative to other fabrics in standard vertical burn tests is shown in Figures 5 and 6. In the char length test (FSTM-191-5903) a lightweight (4.5 oz/yd²)PBI/p-aramid blend had a char length of 0.7 in. whereas an equivalent weight m-aramid/p-aramid blend was 2.4 in. At 6 oz/yd² the aramid blend was still worse at 2 in. Going to the percent consumed test (FSTM-191-5905) the same PBI/p-aramid was 7% consumed whereas the equivalent m-aramid/p-aramid blend was 40% consumed. The 6 oz/yd² m-aramid/p-aramid still had a consumption of 35%.

TABLE 4.

Standard Flame and Fire Resistance Tests

Vertical flammability — (FSTM-191-5903)
Percent consumed — (FSTM-191-5905)
Thermal protective performance (TPP)
Instrumented mannequin

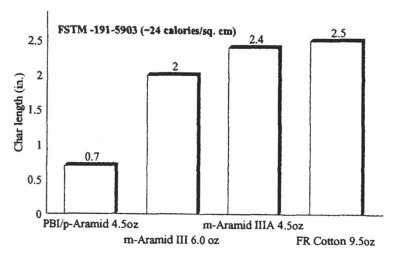

FIGURE 5
Char length of various fabrics after vertical burn test.

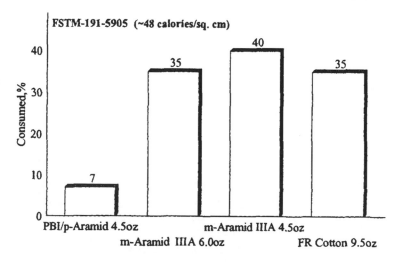

FIGURE 6
The quantity of fabrics consumed after a vertical burn test.

The p-aramid has usually been added at about a 5% level to help prevent flame breakthrough and fabric rupture seen with 100% m-aramid fabrics.

Performance on two other standard tests, i.e., thermal protective performance (TPP) and an instrumented mannequin are shown in Figures 7 and 8. These two tests show similar or only slightly better performance for the PBI/p-aramid blend. Remember, however, that both of these tests represent static situations. This permits a residual char to form an undisturbed insulating layer. The char for the aramid blend is very brittle and in a real life situation it is readily broken by movement, such as when an individual tries to move to get out of the fire. The PBI blend, however, has a flexible char that stays intact and continues to provide protection. This is supported by the information shown in Figure 9 which shows the retention of fabric grab strength after an exposure to a heat flux of 18 cal/cm²/sec. The PBI retained 30–35% of its initial strength whereas the m-aramid/p-aramid and treated acrylic retained less than 5%.

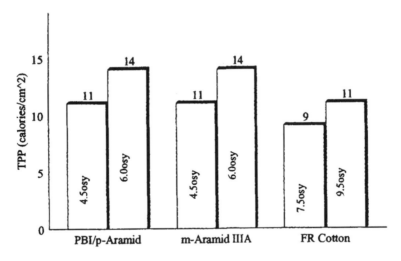

FIGURE 7
The thermal protective performance (TPP) of several fabrics.

Comfort was discussed above as a key element in the selection of protective apparel. Figure 10 shows an apparent temperature index based on dry bulb temperature and relative humidity. Depending on the combination of conditions, the potential level of heat stress can be predicted. Table 5 summarizes average temperature and relative humidity for several cities. For example, an average July day in Atlanta can give an apparent temperature index of 111, sufficient to be in a dangerous heat stroke situation.

Fiber modulus, moisture regain, and fabric construction are all significant factors in the comfort of a garment. Although the p-aramid provides relatively good flame and fire protection, its very high modulus leads to stiff uncomfortable fabrics. A blend of PBI and p-aramid fibers

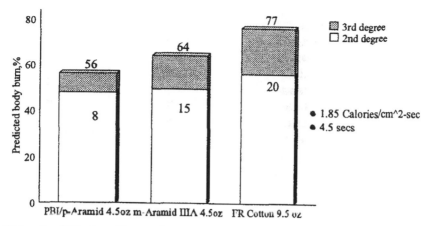

University of Alberta - without underwear.   88% is maximum burn area since hands & feet are unsensored.   7% of available burn area is comprised of unprotected head

FIGURE 8
The performance of several fabrics in an instrumented mannequin burn test.

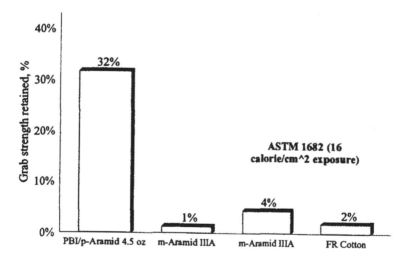

FIGURE 9
Fabric strength remaining after flame exposure.

introduces several positive factors, i.e., the soft hand and high moisture regain of PBI produces a fabric that is comfortable but still has superior flame and fire protection. Figure 11 shows the relative moisture regain of several fabrics. Because PBI has an extremely high moisture regain of 15%, it permits blends with it to have higher regains than other flame resistant fabrics. In previous comfort tests, PBI was perceived to be equal to or better than cotton. The high moisture regain also serves to reduce static electricity buildup, also important to comfort.

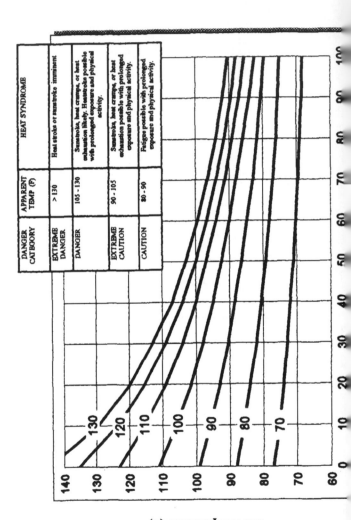

DANGER CATEGORY	APPARENT TEMP (F)	HEAT SYNDROME
EXTREME DANGER	> 130	Heat stroke or sunstroke imminent.
DANGER	105 - 130	Sunstroke, heat cramps, or heat exhaustion likely. Heatstroke possible with prolonged exposure and physical activity.
EXTREME CAUTION	90 - 105	Sunstroke, heat cramps, or heat exhaustion possible with prolonged exposure and physical activity.
CAUTION	80 - 90	Fatigue possible with prolonged exposure and physical activity.

TABLE 5.

Apparent Temperature Index for Various U.S. Cities

City	June	July	August
Atlanta	100	111	113
Charlotte	101	115	114
Chicago	87	91	91
Houston	121	130	130
Phoenix	106	130	125
Tampa	119	121	123
Washington, D.C.	95	109	109

Note: Based on average daily high temperature and relative humidity.

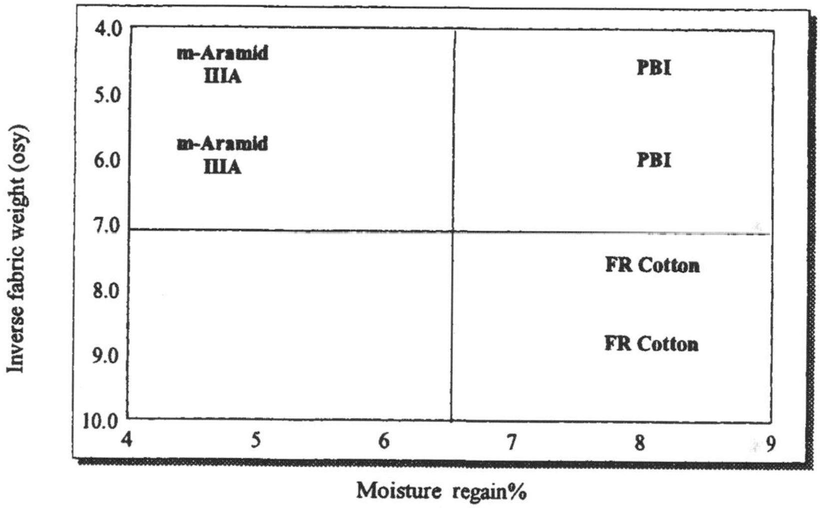

FIGURE 11
The moisture regain of several protective fabrics.

The ability to provide excellent flame protection with lightweight fabrics is an important factor in PBI success. For example, in fire service, more firemen are injured by heat stress than by burns. The recent introduction of a 6 oz/yd^2 50/50 PBI/p-aramid fabric helps lower the garment weight and increase its flexibility to assist in the reduction of heat stress. This is possible because of the exceptional flame resistance of PBI.

VI. EMERGING APPLICATIONS AND TECHNOLOGY

Although the major commercial applications of PBI are as fibers in the protective apparel areas, there are several other areas that offer the potential to capitalize on the unique balance of PBI properties.

A. Electrochemical Applications of PBI

1. Fuel Cells

Significant efforts are now under way to develop fuel cell technology for transportation. For example, the U.S. Department of Transportation has established a national program plan for fuel cells in transportation, driven primarily by environmental considerations. Successful implementation can reduce the dependence on oil imports and reduce the levels of CO, NO_x, and nonmethane organic gas. A major fuel cell candidate is a polymer electrolyte membrane (PEM) fuel cell.

A critical part of the PEM is the solid polymer electrolyte which facilitates the proton exchange from the cathode to the anode and prevents gas diffusion between the two electrodes. Currently, poly(perfluorosulfonic acid) polymer electrolytes are used. These materials generally have a high cost which may be prohibitive for a mass transportation application. In addition they must remain hydrated to retain ionic conductivity, which limits their operating temperature to near 100°C at atmospheric pressure.

One of the more interesting fuel cell candidates employing a polymer electrolyte is the direct methanol–air fuel cell. Its ability to react methanol at the anode and oxygen at the cathode eliminates the need to convert methanol to hydrogen in a reformation process that is then supplied to a hydrogen–oxygen fuel cell. This eliminates the concern for hydrogen storage if the reformation is done off-site or reduces the weight and volume required if the reformation process is done onboard.

At present the performance of direct methanol–air fuel cells is severely limited by the lack of sufficient active catalysts for the methanol anode at temperatures <100°C and the high rate of methanol transport (cross-over) from the anode to the cathode. This results in a loss of efficiency due to the methanol reaction with oxygen and depolarization of the cathode.

One of the challenges for the commercialization of the fuel cell vehicle is the development of a polymer electrolyte with good ionic conductivity (above 10^{-2} s/cm) that is relatively independent of water content; has good thermal, mechanical and chemical stability to the fuel cell environment so that it has an acceptable lifetime; and low methanol permeability at temperatures between 100 and 200°C. It must also be of modest cost, at least an order of magnitude below present poly(perfluorosulfonic acid) polymer prices.

Earlier work with acid-containing PBI showed ionic conductivities that were felt to be too low for effective use in fuel cells.[4] However, recent work at Case Western Reserve University[5] has demonstrated that with the addition of 300 to 500% of acid (e.g., phosphoric or sulfuric) based on the weight of PBI, a highly ionic conducting polymer electrolyte is achieved. Since PBI is a weak basic polymer, it is readily protonated at the tertiary (pyridine) nitrogen in the imidazole ring. The resulting positive charge is delocalized around the imidazole ring. This affinity for acid allows PBI to be doped with acids such as phosphoric and sulfuric, and retention of the acid is quite good even at elevated temperatures.

Phosphoric acid-doped PBI membranes were laboratory tested in a fuel cell at Case Western Reserve University. The cell was operated continuously at 200 mA/cm^2 for 200 h without degradation in performance at 150°C. The maximum power was 0.25 W/cm^2 at 700 mA/cm^2. The gas permeability of acid-doped PBI is considerably lower that of the poly(perfluorosulfonic acid) electrolyte. For example, the undesirable transport of methanol was almost 300 times greater for the poly(perfluorosulfonic acid) than for PBI.

2. Batteries

Applications for PBI in alkaline fuel cells are described in two U.S. Patents assigned to Hughes Aircraft Co. In U.S. Patent 4,217,404, the intrinsically nonwettable polypropylene battery separators, made wettable by treating them with PBI, exhibited excellent thermal and chemical stabilities, showed excellent electrolyte retention, and remained as permeable. In U.S. Patent 4,269,913, an inorganic-organic composite felt for alkaline storage battery separators was bonded with a PBI coating.

In U.S. Patent 4,471,038, assigned to AT&T Bell Laboratories, the addition of PBI into the electrolyte solution increased the capacity and cycle life of cadmium electrodes in nickel–cadmium batteries. PBI, at a concentration of about 1 ppm, apparently decreased the size of the active cadmium species ~1 µm to prevent large crystals (1 to 13 µm) from forming.

B. Polymer Blends

1. Plastic Parts

Although PBI is a thermoplastic, it does not melt and therefore cannot be processed into parts by conventional thermoplastic processing. However, a process was developed[6] using high-pressure sintering techniques to produce fused articles. Products produced by this method are trademarked as Celazole®. Most of the attributes described previously for the fiber are also applicable to the molded polymers, i.e., excellent high temperature performance, chemical stability, nonflammability, etc. However, perhaps the most unique property is the exceptionally high compressive strength. At over 30,000 psi compressive strength, PBI exceeds the performance of any other plastic. Additionally, it maintains its properties at high temperatures. After 500 h at 500°F it had no weight loss or compressive strength reduction. Its compressive strength at 700°F is higher than many plastics at room temperature. Total elastic recovery is obtained at compressive strains of 10% or less.

PBI in 100% form is on the high-cost end of the engineering plastic scale. This is due in part to the ceramic-like processing to produce molded parts, the need for machining to produce final part dimensions, and its specialized nature. However, as with fibers, blending with other polymers offers an opportunity to upgrade their performance by including PBI while retaining processability. For example, blends of PBI and PEEK (poly ether ether ketone) can be thermally processed to provide plastics with enhanced engineering attributes.

2. Polymer Alloy Fibers

Research at the University of Massachusetts discovered that PBI polymer forms homogeneous solutions with a wide range of thermoplastic polyimides.[7] When the solvent is removed, a true polymer alloy, with a distinct and unique glass transition temperature, remains. These alloys permit tailoring of physical properties to select a composition to fit a desired requirement. Since PBI has such an exceptionally high glass transition temperature, it makes an excellent upper end component for alloys, e.g., raising and improving thermal stability, improving chemical resistance. At the high end of the PBI composition range, the polyimide can lower PBI thermal processing requirements. For example, a 75/25 PBI/Ultem® polyetherimide fiber had physical properties quite comparable to regular PBI. However, the glass transition temperature of this alloy was reduced to 375°C from the 425°C of PBI, making the alloy easier to process thermally.

C. Catalysis

PBI can be fabricated into a variety of forms, among them a microporous spherical bead.[1] These beads have recently been demonstrated as a catalyst support in a Wacker-type oxidation of dec-1-ene to methyl ketone.[8] A palladium(II) catalyst supported on a N-cyanoethylated PBI in conjunction with a copper chloride cocatalyst remained active at 120°C. In contrast to homogeneous systems with palladium chloride, no hydrochloric acid was needed to avoid precipitation of palladium metal. The polymer-supported catalyst was recycled seven times with no loss of activity after three recycles.

These investigators have subsequently extended their use of PBI as a catalyst support to immobilize molybdenum(VI) catalysts for the epoxidation of propene.[9] They report that in this Halcon (or Arco) process the use of thermally stable PBI offers a potential for advancing beyond the limitations encountered with less thermally stable supports. They indicate that in comparison with other reported polymer supported systems the PBI support "can only be described as outstanding."

VII. CONCLUSIONS

PBI has recently entered its second decade as a commercial fiber product. The key focus has been to utilize its unique balance of physical properties to provide improved flame and fire protection products. Industry awareness of PBI as a key to enhanced protective apparel has continued to grow so that the familiar PBI gold color has become the symbol of superior flame and fire protection and comfort. A number of additional PBI polymer forms with unique attributes have also been demonstrated, but they are still relatively early in their life cycle.

ACKNOWLEDGMENT

The author is appreciative to Daniel W. Conrad of the Hoechst Celanese PBI Business Unit for information relating to PBI fiber applications.

REFERENCES

1. Buckley, A., Serad, G. A., and Stuetz, D. E., Polybenzimidazoles, in *Encyclopedia of Polymer Science and Engineering*, Vol. 11, 2nd ed., John Wiley & Sons, New York, 1988, 572.
2. Powers, E. J. and Serad, G. A., History and development of polybenzimidazole, in *High Performance Polymers: Their Origin and Development*, Seymour, R. B. and Kirshenbaum, G. S., Eds., Elsevier Science Publishing Co., 1986, 355.
3. Borzo, M., personal communication, 1994.
4. Salzano, F. J., Conductivity of Polybenzimidazole in Steam, *Brookhaven National Laboratories Informal Report BNL 38242*, June 1985.
5. Litt, M., Savinell, R. F., et al., acid doped polybenzimidazoles, a new polymer electrolyte, *J. Electrochem. Soc.*, 141, 146, 1994.
6. Alvarez, E., DiSano, L. P., and Ward, B. C., U.S. Patent 4,912,176, 1990; and Alvarez, E., Blake, R. S., and Ward, B. C., U.S. Patent 4,814,530, 1989.
7. Williams, D. J., Karasz, F. E., and MacKnight, W. J., Recent advances in the studies of aromatic polybenzimidazole/aromatic polyimide blends, *Polym. Prepr. (Am. Chem. Soc., Div. Polym. Chem.)*, 28, no. 54, 1987.
8. Tang, H. G. and Sherrington, D. C., Polymer supported Pd(II) Wacker-type catalysts. II. Application in the oxidation of dec-1-ene, *J. Catal.*, 142, 540, 1993.
9. Miller, M. M. and Sherrington, D. C., Polybenzimidazole-supported molybdenum (VI) propene epoxidation catalyst, *J. Chem Soc., Chem. Commun.*, 55, 1994.

Part IV

HEAT-RESISTANT MATERIALS

Chapter 10

PACKAGING APPLICATIONS FOR HIGH TEMPERATURE PLASTICS

Gene Strupinsky and Aaron L. Brody

CONTENTS

I. Introduction 181

II. Requirements of Products 183

III. Canned Foods... 184
    A. High-Acid Foods Sterilized Outside of the Package ... 185
    B. High-Acid Foods Sterilized Within the Package 190
    C. Low-Acid Foods 191
    D. Aseptic Packaging 197

IV. Heating of Packaged Foods in Ovens 197

V. Microwave Susceptors 200

VI. Conclusion.. 202

I. INTRODUCTION

The principal purpose of packaging is to protect the contents of the container throughout the distribution cycle. If there were no need to move goods from one location to another, there would be no need for packages and package materials. All other functions of packaging, however important they are to consumers, business and society, are secondary to the protection objective.

Product contents must be protected against a continuously hostile environment which is attempting to react with the contents and cause deterioration. Among the elements against which packaging must protect are water, water vapor, oxygen, both visible and ultraviolet light, vibration, impact, static pressure, dirt, microbiological contamination, insects, animals, and today, human intervention. Generally temperature is not included in the enumeration of vectors attacking package contents.

Since the focus of our consultancy is on consumer goods packaging, and especially foods, drugs, personal care items, household chemicals, and other home and hotel/restaurant/institutional disposables, this presentation will address these areas. Concentration will be on foods and beverages because they use nearly 60% of all package materials in terms of both weight and value.

The major materials employed for packaging are paper and paperboard which together constitute about 40% of the total volume. The main use of paperboard is in linerboard and corrugated medium, the basis for corrugated fiberboard cases, the overwhelmingly largest distribution package used in the U.S. Few paperboard packages are called on to resist high temperatures, and thus with the exception of those few this family of package materials is not discussed here. Other materials of consequence in food and beverage packaging include metals such as aluminum and steel, glass, and plastics. One reason for the use of some metals is their heat resistance. Glass is employed for packaging because of its inertness, transparency, and rigidity; and, in some instances, because of its temperature resistance. Each of these two materials represents about 20% of the volume of package materials used in the U.S. In other parts of the world, glass use is higher, and steel is used more than aluminum. Plastic package materials constitute about 10% of the total, with the rate of growth being significantly greater than those of the other materials.

Plastic materials employed for packaging are predominantly the polyethylenes, followed by polypropylene and polyester. Smaller quantities of polyamides commonly known as nylons, polystyrenes, poly(vinyl chloride), ethylene vinyl alcohol, poly(vinylidene chloride) and polyacrylonitriles are also used for food and other consumer goods packaging. Low-density and linear low-density polyethylenes are principally used for flexible films and extrusion coatings on paperboard, paper, other plastics, and aluminum foil.

Low-density polyethylenes and their analogs are desired for their properties of water and water vapor resistance, toughness, and relatively broad and low range of heat sealability temperatures. Their relatively high gas permeabilities have led to their recent application for the packaging of fresh vegetables where a paucity of oxygen from the air could lead to undesirable respiratory anaerobic reactions and off flavors.

High-density polyethylenes find their greatest packaging uses in extrusion blown bottles to contain products with relatively low sensitivity to oxygen, but which require protection against passage of water or water vapor in a tough and flexible container. Major applications are for milk and water, as well as bathroom, household, and automotive chemicals, where they long ago replaced glass, metal, and spiral-wound composite paperboard cans. The properties and good economics of high-density polyethylene bottles have led to bottles and jars fabricated from this plastic being one of the two major categories of plastic bottles in the U.S.

Polypropylene is used mainly for making oriented polypropylene films prized for their clarity and excellent water vapor barrier properties. Such films are used extensively for packaging of salty snacks, cookies, and crackers. Much smaller quantities of polypropylene are incorporated as the inexpensive structural water vapor barrier component of retorted plastic cans. In semirigid form, polypropylene is fabricated into tubs used for hot filling of sauces and cold filling of dairy products.

Poly(ethylene terephthalate) (PET) polyester has been used for film for many years, but as a bottle resin only since the late 1970s. The first major application was for carbonated beverage bottles, but the applications have broadened into bottles for personal care products, and bottles and jars for salad dressings, peanut butter, edible oils, and now aseptically packaged and hot-filled juices and analogous fruit beverages. In crystallized form, polyester is also used for dual ovenable package structures.

The principal applications of nylon are as the thermoforming web in low oxygen packaging of cured meats and cheeses where the polyamides are desired because of their good gas barrier, toughness, and easy thermoforming properties.

Ethylene vinyl alcohol and poly(vinylidene chloride) are relatively expensive resins which are noted for their excellent gas barrier properties; and therefore they are laminated to, coextruded with, or coated on other less expensive plastics which impart the structural and heat-sealing characteristics to the entire structure.

II. REQUIREMENTS OF PRODUCTS

Although each of the thousands of products requiring packaging has a different set of needs, the foods, beverages, and other disposable products addressed in this discussion have only a relatively few. Food products generally need to be protected against water and water vapor transmission, either in or out depending on whether the contents are aqueous or dry. Many food and beverage products require protection against the entry of oxygen which can react with the contents to produce detrimental end products. Some high-fat products require what is often referred to in the industry as grease resistance, but which is really fat and oil resistance. Many products are harmed by light exposure, but the marketing benefits of transparency appear to often outweigh the technical issues.

The reaction rates of food product deteriorations such as enzymatic, microbiological, or biochemical are almost always increased geometrically with increasing temperature above the glass transition point of the product, if frozen; and above 32°F if not frozen. Thus, most products would benefit from distribution at temperatures below the glass transition point for frozen foods and the freezing point for nonfrozen foods. In an ideal world, all packages would have the capacity for perfect thermal insulation or the ability to maintain an appropriate temperature for the distribution time.

Reaction rates refer to three vectors of food and beverage product deterioration: microbiological, enzymatic, and biochemical. Biochemical reactions are slowed by distribution temperature, moisture removal, or reacting the components during processing. Biochemical reactions such as oxidations occur in most foods and are most difficult to control, but, fortunately, are usually not the most critical of deterioration reactions.

Enzymes are biological catalysts with effects that can be devastating to food and beverage products, and which can be obviated by heating the product modestly before or after packaging. Although most enzymes can be destroyed by exposure to 140°F for 1 min, the same effect can be achieved at higher temperatures in lesser times, with fewer adverse effects on the product itself. For example, exposure at 270°F can achieve enzyme destruction in less than 10 s. Most high temperature enzyme destruction is performed outside of the package because of heat transfer problems, but one of the reasons is a limited availability of package materials capable of withstanding the temperatures required to achieve the requisite temperature within the package itself.

The most critical deterioration vector is microbiological, for which effects can be mitigated by distribution at temperatures below the freezing point or about 26°F for frozen foods, or slowed by distribution at temperatures close to 32°F if distributed at refrigerated temperatures.

In many instances, microorganisms are destroyed by heat, with those capable of growing under high-acid conditions destroyed by temperatures below 212°F within a few minutes either in or out of the package. Microorganisms capable of growing at a pH above 4.5 under low oxygen conditions are often potentially pathogenic and therefore must, by law and common sense, be destroyed. Thermal methods to destroy heat-resistant anaerobic pathogens require exposure to prescribed temperature-time protocols which are in the vicinity of 250°F for 1 min at the slowest heating point in the product. Reduction of time required to heat and thus reduction in adverse excess thermal input to the product may be achieved by increasing the temperature. One result of thermal sterilization is that enzymes are usually destroyed by the treatment.

III. CANNED FOODS

Obviously, the time–temperature profile of the slowest heating element of a food within a package depends on the time–temperature profile imparted to the exterior of the product which, in most instances is the package, which translates into significantly higher temperatures for longer periods for the package. Both metal and glass packages and their closures are capable of withstanding the 250°F and even higher for up to 1 h, or occasionally even longer. Among the challenges for plastic materials and structures has been the ability to withstand the time–temperature profiles required for thermal microbiological destruction or sterilization.

As implied, in-package thermal sterilization may be divided into three separate types:

- High-acid foods
  - Sterilized outside the package and filled hot at temperatures of 170 to 210°F
  - Sterilized in the package to achieve internal temperatures of up to 210°F
- Low-acid foods
  - Sterilized inside the package at temperatures in the 250°F or higher range

Another technology, called aseptic packaging, involves sterilizing the product and package independently of each other and bringing the two together under sterile conditions. When the product is fluid, it is relatively easy to thermally sterilize it outside of the package and to sometimes achieve a relatively high quality by virtue of the lower total thermal input. Many fruit beverages are treated in this manner with good commercial results. When the product contains particulates or is low acid, it is much more difficult to thermally sterilize the product outside of the package and to transfer it under sterile conditions into a presterilized package.

Each of these thermal sterilization technologies will be discussed in the perspective of the package temperature resistance requirements.

A. High-Acid Foods Sterilized Outside of the Package

To ensure sterility, it is necessary to sterilize the interior of the package. The most common means is filling at product sterilization temperatures of 170 or 210°F, depending on the product. At these temperatures any microorganisms on the interior of the package are destroyed. Without dealing with issues other than temperature, it is evident that the traditional metal can and glass bottle or jar are quite capable of resisting these temperatures, not collapsing under the vacuums formed as a result of later condensation and of excluding air which would otherwise either bring in contaminating microorganisms and/or oxygen. The introduction of plastic material structures was proposed to achieve lighter weights, lower cost, design flexibility, or direct in-package microwaveability. Among the challenges for the materials are temperature resistance which, by itself, is not difficult to attain. The package material must axiomatically be safe from a regulatory standpoint, meaning nothing detrimental from the package material enters the product at the processing temperature, or under ambient temperature distribution. Further, the package material must have the oxygen, water vapor, water, and fat barrier characteristics to permit the requisite shelf life of up to 6 months for most products. Of increasing importance to food packagers is the assurance that the package material does not remove desirable flavor attributes from the product during the thermal sterilization treatment or later during distribution at ambient temperatures. The package must be hermetically sealable to guarantee the exclusion of contaminating microorganisms and the closure must be accomplished by a practical packaging machine. All of this must be done at a package cost that competes effectively with traditional and already low-cost glass and metal. For a 1-lb food can, this is perhaps no higher than $0.20.

Hot filling is now successfully performed commercially for pumpable products such as tomato sauces and ketchup in plastic and plastic-aluminum foil pouches in sizes from 1 oz. up to 3 gal. Materials typically employed include linear low-density polyethylene on the interior as a broad-range heat sealant and nylon 6 on the exterior as a principal oxygen barrier and tough protector against impact, abrasion, pinholes, and stretching, especially during the stresses of heat and pressure during filling. If aluminum foil is present, it is the principal barrier and must be protected. The interior heat sealant serves as a cushion against hydraulic shock during distribution. Pouches may be preformed and opened and filled, but more often are fabricated in-line from roll stock laminations or coextrusions, with the hot filling in the same vertical form/fill/seal machine. In the larger sizes, this type of package replaces the large number 10, 6- to 7-lb capacity, hotel/restaurant/institutional steel food can with a much lower cost, lighter weight and easier to open, dispense-from and dispose-of package.

Performance under elevated temperatures is essential during hot-filling operations to ensure package integrity. The pouch is initially formed, usually with face-to-face polyethylene-to-polyethylene fusion heat seals at temperatures of about 300°F, immediately prior to filling and thus must set within less than 1 s before being stressed with fluid at up to 210°F, a condition which persists for several minutes. The open top is then heat sealed, and almost immediately the hot fluid is in contact with this newly formed heat seal. Thus, the sealant material must not only resist the temperature but also the pressure of hot liquid for several minutes until exterior coolant is applied and begins to reduce the temperature. The interior sealant must possess a property called hot tack, or the ability to retain its physical integrity under stress at temperatures in the 212°F range. To date, linear low-density polyethylene has been satisfactory, although its performance has not been more than marginal, and great care must be taken during operations since there is little latitude for deviations. It has been suggested that among the newer metallocene-catalyzed polyethylenes, better performing sealants will be found and thus afford the possibility of expanding the range of hot filling into flexible pouches.

This category of product sterilization also encompasses fruit beverages and other fluids packaged in metal cans; glass jars; and semirigid plastic bottles, jars, and even cups. Among the problems is that after cooling to ambient temperatures, the steam in the headspaces of these packages condenses, leaving a partial vacuum within the container. With rigid metal and glass, the differential pressure is of relatively little consequence; however, with semirigid plastic bottles and cups, the wall strength can be so low that they tend to collapse, resulting in unattractive appearing packages. This problem is being overcome by structural design of both bottles and cups, for example, by incorporating ribbing into the walls or increasing the gauge, a cost-adding alternative.

Temperature resistance of the plastic is not as easily overcome. If the product need not be seen, then high temperature resistant combinations such as coextrusions are often possible, but when one key is to permit product visibility, the current options are limiting. Products such as ketchup have been hot filled into coextrusion blow-molded bottles fabricated from polypropylene as the moisture-barrier and temperature-resistant component, and ethylene vinyl alcohol as the oxygen barrier. Since temperatures involved are only in the 170°F range, thermal resistance has not been a serious issue. As the marketing desire for better appearance and environmentalist solid waste issues surfaced, however, multilayer and opaque bottles became less desirable. Improving the surface through polished blow molds and incorporating clarifying agents in the polypropylene helped to address the marketing desires, but not all the environmentalist concerns. The application of polyester with its very marginal heat resistance in the 170°F range has been employed commercially, but, in this instance, the product is being packaged under aseptic conditions in which the temperature limits of the bottle plastic are not being challenged. As a significant comment, the current commercial material is a proprietary coinjection of polyester and small quantities of ethylene vinyl alcohol, meaning that there is a need for a transparent bottle plastic for ketchup that has higher heat resistance than currently available commercially for most packers of the product.

During the 1980s, considerable developmental effort was directed toward increasing the temperature resistance of polyester bottles through partial crystallization of the material to retain its desired transparency and toughness. Because of the need to hot fill at temperatures approaching 185°F, crystallization appeared to be the route to increasing temperature resistance. The glass transition temperature (T_g) of amorphous polyester is in the range of 158°F, with the glass transition temperature of the oriented side walls only slightly higher. When the standard untreated polyester bottle is heated, residual strains relax and the structure tends to revert to its original form, thus shrinking and distorting the bottle. The (T_g) increases with crystallinity. Orientation in the bottle walls increases crystallinity to about 28% which increases the T_g to about 163°F. Obviously, increases in T_g as a result of orientation do not occur in the bottle neck or base. The biaxial orientation is required to maintain the bottle clarity since the unoriented crystals would be milky white.

Applying heat to polyester can increase its heat resistance by increasing its crystallinity. At 50% crystallinity, the T_g can be as high as 300°F; and at peak crystallinity, the T_g can reach 330°F, but the appearance is almost opaque (satisfactory for dual-ovenable trays but not for beverage bottles). At 32% crystallinity, the T_g is raised to about 185°F, which is marginally satisfactory for much fruit beverage hot filling. Crystallinity is increased by heat treating the bottle during the blow-molding phase of fabrication. The process, sometimes imprecisely called heat setting, involves heating the mold to about 200 to 220°F while maintaining the internal pressure for up to 20 s. During this period, molded-in stresses are relaxed and the molecules are better aligned. Although this is current commercial practice, the system has drawbacks: equipment to control the mold temperature is difficult because of the special cooling required, and the maximum temperature of the bottles is well below that used by glass packers. Many products require higher temperatures, and some of the very high-volume products such as tomato sauces, pastes, and concentrates require temperatures of up to 210°F to ensure sterility. Technologies from Yoshino (Japan) now with Johnson Controls, Continental PET Technologies, Constar, Graham, and others are achieving the 185°F level; and millions of bottles in sizes ranging from 16 to 128 oz are commercially employed to deliver fruit juices, fruit juice drinks, and isotonic drinks.

Increasing T_g by increasing crystallization has been taking place in incremental steps to achieve 195°F, seemingly a small step, but nevertheless important for fruit beverage packers. Krupp Corpoplast and Nissei both have developed systems to increase crystallinity to 35% by blowing into a very hot 300°F mold and cooling the interior before demolding in the case of Corpoplast. The Nissei oven blow process sends the partially annealed bottle into a machine where it is heated before it is reblown.

Both Krupp Corpoplast and Sidel have developed means to incrementally increase crystallinity further to permit longer exposure to 190°F by the bottle during hot filling, i.e., enter the phase by which the product may be in-package pasteurized. In the Sidel process, the injection-molded preform is reheated and overblown but into a larger size than the final bottle. The oversized bottle is shrunk by exposure to 400°F after which the bottle is blown back to normal desired size. The Krupp Corpoplast process is similar using another version of two blowing stations on their bottle fabrication machine. It is more than interesting that the means to increase temperature resistance involve injection molding and sequential blow molding under abnormal conditions, and still the temperature resistance of the polyester has not reached 210°F at which temperature an entirely new category of products such as tomato juices could be conveniently packaged.

Obviation of the temperature resistance issue has been accomplished by aseptic packaging, but this is not entirely satisfactory as will be discussed below.

The results of the mechanical approach to increasing temperature resistance of polyester lead to the inevitable question of the employment of alternative plastic resins. The identification of a plastic resin that meets all the requirements of all the disciplines has not yielded a suitable material to date. Reiterating the requirements, the final plastic structure must have oxygen and water vapor barrier resistance, exhibit no flavor interactions with the contents, be acceptable to regulatory agencies, be inexpensive since its major competitor is glass, be transparent in distribution, be relatively easy to fabricate, and be impact resistant as well as have sufficient temperature resistance to withstand up to 212°F. Some plastics such as the polycarbonates have almost no barrier properties although one commercial venture by Graham Engineering with General Electric led to coextrusion blow molding of polycarbonate with a nylon core barrier plastic which turned out to be far too expensive for commercial practice because of the polycarbonate component.

Other entries which lacked all of the requisite properties included Kamax acrylic imides from Rohm and Haas and the very high temperature resistant but very expensive U-polymer polyarylates from Unitika (Japan) which were coextruded with polyester in blown bottles.

The most promising contender at this time is poly(ethylene naphthalate) polyester, now referred to commonly as PEN. This resin has a gas barrier five times that of PET, water vapor barrier three times better than that of PET, and a glass transition temperature of 240°F or about 70°F higher than PET. PEN also has excellent mechanical properties, particularly its modulus which is 50% higher than that of PET. Its tensile strength is also 33% higher than that of PET. PEN thus has the potential in applications for packaging of high-acid products requiring the higher hot fill temperatures and for products requiring postfill pasteurization at temperatures in excess of 212°F. Among the products which might be packaged in PEN are tomato sauces and purees, baby juices, pickles, and fruits. (Another intriguing market not directly related to temperature resistance is for small, 16-oz and below bottles of carbonated beverages with a shelf life that has been limited by the gas permeability of PET.) Development technologists place the hot fill temperature of PEN at about 212°F.

PEN is produced from naphthalene dicarboxylate ester (NDC) or 2,6 naphthalene dicarboxylic acid (NDA) (Eastman source) as the raw material. Amoco has built a plant to manufacture NDC from orthoxylene which will be linked to a facility to make the dimethyl ester for use in making the PEN. PEN has recently received clearance from U.S. regulatory authorities for contact with certain foods. In Europe, some commercial bottles and jars incorporating PEN have been introduced.

Currently pricing for PEN is projected into the range of $1.50/lb which is about 2 times the price of bottle grade polyester, a major obstacle.

It appears that one means of deriving the benefits of PEN at more acceptable economics are to blend or alloy the PEN with PET, a development that was commercialized for salad dressing and marmalade by Skillpack in The Netherlands. The PEN content was less than 50% of the blend, with some estimates as low as 20%. Price of the jars was about 50% above the equivalent PET jars. In the realm of food packaging, a 50% upcharge on package material costs is generally out of range unless some clear functional or marketing benefit is achieved. The question here will be do consumers care if their jams or tomato sauces are in clear plastic enough to pay a premium for the package.

Another alternative to take advantage of PEN properties at acceptable cost would be to coinject or coinjection blow mold PEN with PET.

One conclusion from this discussion on polyester and its PEN analog is that the definitive heat-resistant plastic for hot filling has not yet been commercialized. Also it has not been developed sufficiently to have been introduced or announced. There is every reason to believe that a market exists for such a plastic material when, and if, it is developed.

One aspect of this hypothetical discussion that should be of interest to polymer developers is that in previous developments in plastic bottle packaging, packagers have demonstrated a willingness to compromise their specifications to enable them to apply the new plastic. The introduction of polyester carbonated beverage bottles was assisted by bottlers reducing their shelf life requirements. Some fruit beverage packers appear to have relaxed their temperature–time protocols for sterility for their products to accommodate the polyester. Thus, a precedent exists for the possibility of packer compromises to permit the commercial application of new heat-resistant plastic packaging. The ability to vary the protocol derives from the fact that the contents are high acid and therefore are not vulnerable to pathogenic microbiological growth. The willingness to alter the protocol comes from the packer's desire to gain the enhanced quality from reduced time–temperature exposure of the product.

B. High-Acid Foods Sterilized Within the Package

Most products failing into this category are packaged in cans or glass jars, being first hot filled and then closed and heated after closing at temperatures that conveniently are 212°F, which is sufficient to achieve sterility for high-acid products such as solid fruit and tomato products. There is relatively little difference between the requirements for plastics for postfill pasteurization and hot filling, and so much of the discussion above on hot filling is applicable with the major exception that the temperature requirement is for 212°F and not for some lower value. Further, the exposure to the temperature is for considerably longer than the momentary, perhaps 10-s, exposure in hot filling. The plastic must be able to withstand up to 30 min of 212°F with intact walls and heat seals, if employed. Even if mechanical closures are employed as in mechanical seaming with a metal end, the seam area must not soften under stress and thus cause permanent distortion to the seam and create passageways for microbial contamination.

The notion of increasing the exterior temperature to increase the rate of heat transfer has been advanced, but if this were a valid food processing unit operation, it would have long since been practiced since glass and metal have been the package materials of choice to date.

Very few commercial examples of high-acid foods that have been pasteurized exist. During the mid-1980s, Metal Box in the U.K. tested a Stepcan polyester tubular extrusion can body with seamed metal ends containing fruit pieces in syrup, obviously hot filled and postfill pasteurized. The ambient temperature shelf life of the contents was too short for any realistic commercial application. Since the polyester body was clear, the maximum temperature must have been in the 170°F range; and thus the temperature of pasteurization must have been below this temperature. Also during the 1980s, another highly publicized thrust involved Continental Can and its injection blow-molded polyester Thermopet can now being used for packaging of tennis balls. This two-piece can with a single-seamed metal end was hot filled with spaghetti sauce and probably postfill pasteurized. This package might have been heat treated, but the maximum temperature could not have exceeded 180°F; therefore, the temperature protocol must have been a very long time, leading to marginal initial quality. This package was also withdrawn as a food package, again signaling the need for higher heat-resistant plastics for thermally processed food packaging.

Throughout the 1990s, several Japanese plastic package converters such as Toyo Seikan have been supplying thermoformed cups fabricated from coextrusions of clarified polypropylene and ethylene vinyl alcohol (EVOH). The contents are cut fruit in syrup which is hot filled and, in many instances, postfill pasteurized at temperatures slightly in excess of 212°F. The EVOH component of the coextrusion imparts the oxygen barrier and the polypropylene, the water vapor barrier and the thermal resistance. Clarified polypropylene offers good but not polyester-like contact transparency. The product is often distributed under refrigerated conditions, helping the quality retention period; however, since the product is sterile as a result of the thermal process, no microbiological shelf life issues are present. Since the temperature is raised above the 212°F point of boiling water, the plastic cup must be made of materials capable of resisting the elevated temperature, and polypropylene fits this requirement.

C. Low-Acid Foods

Low-acid foods such as meats, vegetables, and pastas present a totally different problem for packaging. As indicated above, thermal stabilization requires temperatures well above 240°F to ensure that the slowest heating point within the package has reached the requisite time–temperature protocol. Required by law and regulation, this thermal profile is necessary to avoid the growth of anaerobic pathogenic microorganisms which are capable of reproducing in the low-acid anaerobic environment within a hermetically sealed package. For almost a century, cans and glass jars have been effective temperature and pressure vessels for thermal sterilization of low-acid foods. Prior to that time, the procedures were practiced with no understanding of the scientific principles involved.

Almost since the development of the first plastics, attempts have been made to employ thermoplastic package structures for cost and weight reduction and for other reasons that are enumerated below. The basics for sterilization are, however, the same regardless of the package material or structure. The internal temperature of the slowest heating point in the package must reach the requisite temperature which is always in the 240 to 270°F range. The rate of heat penetration varies with the size and shape of the package, the composition of the package since each package material has a different heat transfer rate, the product since each product has a different thermal transfer characteristic, the external temperature, the motion imparted to the package and its contents during sterilization processing, and other variables. With only a few exceptions, sterilization of low-acid foods requires that the package become essentially a pressure vessel and that sterilization be performed in a pressure vessel. Regardless of the use of pressurized steam as the external heating medium, if a closed package is heated, the package becomes a pressure vessel. The exceptions of using an open flame or superheated steam as the heating source or open top heating are infrequent so as to be omitted from this discussion.

Since the interior contents are largely water, as soon as the temperature reaches above about 212°F, significant steam pressure begins to build within the package and a balance must be maintained between the interior and exterior pressures to avoid distortions and explosions. With glass containers, the pressure differential is even more critical because the closure is generally held on primarily by the internal vacuum; and when steam within the jar increases, the closure has no force retaining it unless the exterior pressure is increased to counter the interior pressure.

The greater the thermal experience by the product, the lower the food content quality; and thus the time–temperature profile is computed to achieve the required sterilization effect and the temperature is reduced as rapidly as possible, maintaining the internal-external pressure balance required.

All of these variables and others not enumerated here lead to the need that any plastic structure for sterile low-acid foods must be able to resist temperatures above 250°F and internal pressures that can exceed several inches of water, as well as, being safe when in contact with the food at the elevated temperatures and having sufficiently low cost to compete effectively with steel, aluminum, or glass. As with other potential applications of plastic for food packaging, this requires processor, distribution, or consumer benefits other than economics such as impact resistance to warrant development investment.

The earliest venture into plastic packaging for low-acid foods was the retort pouch started during the 1950s on the premise that the higher surface-to-volume ratio of the flat pouch would permit faster heat penetration and hence better quality food. Further, because of the lower mass of package material, the flexible pouch generally should have a lower cost than a metal can. As the military entered as an advocate, the significantly lighter weight and ability to be placed in a soldier's pocket became factors. The first retort pouches were composed of aluminum foil laminations with interior heat sealants and exterior plastics to protect the aluminum foil which constituted the main barrier. After many years of development both for the military and for European and American commercial applications, the structure of choice is an exterior of biaxially oriented heat set polyester film, a core of aluminum foil, and an interior of cast polypropylene which functions as a fusion heat sealant. Each of the components including the polyurethane adhesives linking the main materials has individual temperature resistance in excess of the 250°F employed to thermally sterilize the contents.

The retort pouch is now a constituent of U.S. military rations, and enjoys some success in Japan where the relatively slow speed equipment is acceptable to food processors in that country. Attempts to obviate the use of aluminum foil in the structure have not been highly successful. Aluminum foil suffers from pinholing and cracking at the interfaces of heat seal and body wall areas and is fragile under repeated stresses. Several plastic materials including liquid crystals have been proposed as aluminum foil substitutes; however, to date, the temperature resistance, economics, and barrier properties of aluminum have not been surpassed by any plastic structure.

During the 1980s, the search intensified as a result of the worldwide spread of microwave ovens in which the aluminum foil is a microwave energy block. Marketers wanted a retort pouch material to permit the product to be reheated in microwave ovens. The closest approximation to date has been silica-coated polyester film which remains too expensive and only available in limited pilot plant quantities.

Efforts to use plastics as metal can substitutes during the 1960s and 1970s led to the application of polypropylene for can bodies mainly because of its temperature resistance up to nearly 270°F, which is satisfactory but not outstanding. Because polypropylene is a poor oxygen barrier, the plastic was always laminated to aluminum foil in insert injection molding as with the Italian Star Pack or in extrusion lamination operations such as with the Swedish Letpack. The Star Pack was hermetically closed with a mechanical seam; the Letpack was sealed by a combination of mechanical clinching and heat fusing the polypropylene to itself on the body of the can. Counterpressure was used when sterilizing the cans. Further, temperature control in the retorts was greater than for metal can retorts because of the temperature sensitivity of the plastic as contrasted to that of steel. The costs of these two can structures proved to be higher than those of metal cans; and with no other visible benefit to the packer or the consumer, the developments were terminated — Letpack in 1994. The ventures demonstrated again that if an effective, safe, and economical high temperature (up to 300°F), resistant plastic were commercially available, the retorted package might have a viable market.

The advent of the microwave oven led to a major thought process change: frozen dinner platters in the 1960s became the package, the heating vessel, and the serving and consuming dish. If canned foods were in a microwaveable package, then the convenience of frozen dinner platters could be delivered in ambient temperature shelf stable form: a take-it-anywhere package. Development had begun during that period on multilayer polypropylene–ethylene vinyl alcohol structures with an objective of being metal can substitutes. At the outset, the cans produced were just that, metal can clones: formed straight side wall bodies hermetically closed with mechanically seamed metal ends. Consumers could not detect any difference between the conventional metal can and the new plastic can, and the processor had the added problems of the thermal sensitivities of the plastic and the difficulty in ensuring a hermetic seal with conventional seaming equipment. Further, the plastic multilayer cans were, of course, more expensive than steel cans.

During the mid-1980s, the concept of the plastic can being the ambient temperature shelf stable version of the frozen dinner platter emerged with the development and introduction of the bucket type can/container. With the application of an easy open full panel pull-off aluminum end plus a heat-resistant high-density polyethylene snap-on overcap to use in the microwave oven to contain the hot food plus a foamed polystyrene full body wraparound label to function as a graphic communication medium and a thermal insulator for the consumer, the concept of microwave reheating of canned foods began. The can body was fabricated from multilayer structures of polypropylene and ethylene vinyl alcohol which in their various manifestations provided satisfactory ambient temperature shelf life.

Persons who initiated the project with the several pioneering supplier-converters have often reflected that the selection of the plastic materials was the easiest part of the development. Although hardly perfect, polypropylene had sufficiently high temperature resistance for both the retorting and, obviously, the microwave reheating. Issues still existed such as moisture sensitivity of the ethylene vinyl alcohol gas barrier, individual layer recovery and reuse of the scrap from thermoforming or coinjection blow molding when either of these fabrication methods was used, heat stresses in the polypropylene that developed into microcracks in the seaming perimeter during closing operations, objections from vocal environmentalists concerning the several different materials used that could not be separated for recycling, flavor interactions of food contents and plastic, and so on. Probably the most intriguing of the problems was the lower thermal conductivity of the plastic body and the consequent need to adjust the thermal process to the reduced heat penetration rate.

Despite resolving most of the technical problems, many of which, as can be seen, revolved around the temperature sensitivities of the plastic materials at various stages of the operations, the product has not succeeded after a very promising start. The reasons can be summarized into one fact that the plastics resin suppliers and fabricators did not reckon in the development: the product as eaten did not taste good enough to the consumer. If as much human resources that addressed the package structure and performance were invested in the food formulation and thermal experience, the package could have had several orders of magnitude greater sales than the current 200 million units in 1994.

An interesting variant on the original polypropylene–EVOH can was developed during the early 1990s with the expectation of superior heat resistance; a monolayer and hence lower cost and easier-to-recycle structure; and less flavor interaction. Crystallized polyester could exhibit temperature resistance of up to 300°F and offer much better oxygen and water vapor barrier than amorphous polyester, even approaching that of the EVOH–polypropylene in what were then conventional buckets. Since the cans were opaque, the crystallinity was not an issue relative to optics or appearance. Among the problems were that the crystalline polyester was much more fragile than amorphous polyester or even polypropylene, and thus far more microcracks developed in the seam perimeter than with the softer polypropylene leading to much higher risk of microbiological contamination. A few crystallized polyester buckets reached the commercial market but this program has been slowed, if not terminated as a result of the disappointing market results.

Paralleling the development of the retortable plastic can and its bucket form was the development of the retort plastic tray, a flat heat-sealed version of the bucket. Closure is usually with a flexible structure heat sealed to a flange of the tray. The notion here is analogous to that of the retort pouch, which also began with this same premise. The reality of the reduced thermal input to achieve the requisite sterility was, and is, different from the theory. Concerned with the unusual-for-the-objective shape led to overprocessing to guarantee the sterility of the contents. More important, however, is the ability to produce a hermetic heat seal with a flexible closure that can be opened without resorting to an implement. One of the reasons for this package is convenience for the consumer, an attribute not achieved with a difficult-to-open package, as the persons who test marketed the Plastigon Campbell Soup package during the mid-1980s can confirm. This soup bowl was thermoformed from polypropylene–EVOH, eventually with wide flanges to which a flexible membrane was heat sealed with nearly fusionlike persistence. That the developers had problems is indicated by the narrow flanges on an early version, forcing consumers to grasp the microwave-heated package by the hot body. In a later version, winged flanges permitted consumers to hold the hot package by the relatively cool outer wings of the flange. Nevertheless, the conflict between the imperative of hermetic sealing and the desire for easy opening was never resolved by this package which was withdrawn.

Because the Plastigon was a round tray, the thermoforming-induced stresses on the body during retorting did not manifest themselves strongly as stressing the heat seal–body interface. When the shape went to rectangular to fit perceived consumer desires, the thermal stresses strained the heat seals which, as cannot be overemphasized, must remain hermetic throughout thermal processing at up to 260°F and during subsequent distribution. This problem was overcome by those who employed trays fabricated by melt-to-mold technologies. Even here, however, the upper temperature of retorting was, and is, limited by the polypropylene component. The issue of heat sealing is somewhat resolved today by one of several technologies: Continental uses fusion sealing of a flexible closure coextrusion of polypropylenes which fractures at the interface of the coextrusion rather than at the interface of the closure and the tray body; and newly developed thermoplastic adhesives are used, which with careful temperature/time/pressure control can effect peelable hermetic seals.

One alternative attempted, even before its use in buckets, was the use of crystallized polyester for the tray bodies. As with the bucket situation, the idea was that the crystallized polyester has both very high temperature resistance and good, although not outstanding, oxygen barrier characteristics. As with the buckets, however, the crystallinity to achieve the temperature resistance did not help the physical impact resistance of the tray structure.

These examples of the realities of plastic retort can, bucket, tray, etc. performance illustrate the major issue: high temperature resistance alone is not nearly enough for the complex and often conflicting needs of food packages that are to be heated once or more than once during their existence. The ideal plastic would have all the attributes displayed by steel plus aluminum plus glass, without all their drawbacks. The temperature resistance must be accompanied by inertness both in flavor and regulatory concerns; barrier; absence of heat-induced stresses; no distortion; ease of hermetic closure, one of the most difficult of all objectives; ease of fabrication; low cost; and appearance. Although many engineering polymers meet the high temperature requirement, they have not met all of the other tests, as can be confirmed by reexamining the history of retort cans, pouches, trays, etc., which is sprinkled with the debris and financial losses of numerous polymers and polymer combinations proposed simply because of their temperature resistance. In our experience, too many developers coming from the polymer side almost ignore the food package requirements such as those presented here. Items like hermetic seals, no flavor interaction, oxygen barrier at elevated temperatures, water vapor barrier, etc. have been among the properties dismissed as less important than temperature resistance by polymer engineers.

D. Aseptic Packaging

As indicated above, aseptic packaging appears to require less heat resistance on the part of package materials than hot filling or postfill pasteurization or retorting because each component is independently sterilized. In fact, one of the benefits cited for aseptic packaging is the lower thermal input imparted to both the product and the package materials, resulting in better quality product and the ability to use almost any package material. In general, these are facts; and for aseptically packaged high-acid products, the package material temperature requirements are usually lower than for hot fill and/or postfill pasteurization. Some commercially available systems, however, employ the same chemical or physical sterilization procedures for high- and low-acid product package materials. These processes on either paperboard/aluminum foil/plastic laminations or all-plastic sheet formed in-line or preformed plastic cups or tubs, translates into exposure to 35% hydrogen peroxide at temperatures of nearly 200°F, exposure to steam at temperatures of well over 300°F, or even if possible to superheated steam at temperatures of over 500°F on the Dole system, all for more than an instant.

Thus, as with the more conventional food thermal stabilization processes, the temperature resistance requirements of aseptic package materials often are not at all different and may have even more nonthermal property needs. As with all of the food package situations cited here, it is critical to incorporate all of the food package needs in order to even consider the potential for commercial application.

IV. HEATING OF PACKAGED FOODS IN OVENS

With the shift in marketing of foods toward packaged foods for total convenience, in which the package is the heating and eating device, the desire has been for packages that can be heated directly, exactly as were the aluminum foil trays of the frozen dinner platters and pot pies. Aluminum foil served this objective admirably until the microwave oven became a household appliance during the 1980s. Because metal is a microwave energy shield and because almost every home had a microwave oven, package development turned to structures that could function in both the conventional and microwave heating environments.

This search suggested to some packagers and their developers the use of very high temperature resistant plastics for conventional convection, conduction, and radiation ovens in which broiling, baking, and grilling operations take place. Temperatures as high as 500°F are sometimes reached and thus the package materials would have to be able to resist this environment. Very few food packager/marketers have asked for this temperature range to replace aluminum, steel, glass, etc. If someone has such a polymer, it might be proposed; however, as with the plastic requirements for canned foods, both total function including cost, barrier, and FDA requirements would enter into the equation. This requirement prevails whether the material is employed as a package material or not, since FDA has, probably rightly, claimed jurisdiction over cooking utensils.

The concept of dual ovenable packages began in the 1970s to try to accommodate the totally different temperature requirements of the microwave and the conventional oven. Food within a microwave oven heats by interaction of the microwave energy with polar molecules, mainly water in the food. With microwave energy penetration into the mass of food, the food heats to a maximum temperature of 212°F until the water has been totally evaporated. Further, there is a comparatively low-temperature differential between the surface and the interior of the food, leading to an absence of surface browning and crisping effects often regarded as highly desirable. Because the microwave energy heats the food directly, the oven environment generally is at ambient or slightly higher temperature. Food packages for microwave heating only therefore have temperature requirements of up to 212°F, although there are some exceptions. This is the reason that polypropylene could be applied for packages for retort buckets that were to be reheated in the microwave oven.

During the early phases of domestic microwave heating, the market appeared to demand packages that could be used in both the microwave and the conventional oven. Several different approaches were tried, some of which were simply to employ microwave transparent high temperature resistant engineering plastics such as polysulfones. The most successful of these plastics both technically and commercially were thermoset compression-molded mineral-filled unsaturated polyesters which could be heated up to 400°F. With contents of water-based foods, proximity of the package material to the product exerted a cooling effect so that even in conventional ovens in which some sectors might have reached temperatures in excess of 400°F, the package structure could withstand the thermal effect.

Among the variables that were addressed during the development of this package structure were the temperature limits which could be moderated by the presence of the food contents, odors that were emitted from the plastic when the material was not properly cured prior to use, and the potential interaction of the mineral and other fillers with the food. Again, temperature resistance of the polymer was only one aspect of the total development. This package was used for Armour Dinner Classics, the first of the many frozen dinner platters aimed directly at the dual-ovenable market. The relatively high cost of the package led to its replacement several years later by crystallized polyester which, despite lesser temperature resistance, proved sufficiently satisfactory to be used commercially to this day even though it does not have the feel of expensive dinnerware.

The earliest technology applied for dual-ovenable packaging (beyond glass) was coating paperboard with high temperature resistant polymers under the premise that the paperboard would not begin to char until the temperature reached well over 450°F, and that the paperboard would function as a structural member while the internal coating would protect the paperboard from the product. The presence of moisture in the food, and incidentally in the paperboard, would ensure that the package temperature would remain close to 212°F until the product moisture was evaporated. The premise appears to have been well founded since most, but not all, paperboard did not char in conventional ovens. The system had sufficient merit that it remains in widespread commercial use; although today, the major use, by far, is for microwave-only applications.

The interior coatings of the paperboard, and of film laminated molded pulp, also an effective structural member, were present not as a heat-resistant structural component. Rather, the coating, or film, is present to protect the paperboard, which is very sensitive to water and fat, against these two components of the food. Because of the high temperature (above 200°F) exposure, the internal coating had to resist these temperatures. Thus, common low-density polyethylene coatings which serve well for nonthermal environments are precluded. The first coating system applied was from the acrylic family, but these materials proved to emit undesirable odors when exposed to elevated temperatures in the conventional oven. These were replaced by polyester which displays the requisite temperature resistance in the paperboard structure and also has excellent moisture and fat resistance as well as being virtually inert to food interaction even at higher temperatures.

Polyester-coated paperboard trays are, of course, quite acceptable for microwave oven use, but are hardly perfect for conventional oven use. The polyester itself is not damaged since it crystallizes into a higher temperature-resistant form at elevated temperature. Rather, the paperboard itself tends to char in areas of the package that might not be in contact or close proximity with the contained food. This flaw led many packagers to consideration of all-plastic package structures. During the 1980s, more than 20 different plastic offerings were made by polymer suppliers, mainly based on the temperature resistance of the material, and with less than complete consideration given to the many other variables involved in food packaging.

Among the many plastic materials proposed, some of which became commercial if only for a brief period, were polyarylates, nylon 11, polycarbonate/nylon, polycarbonate, polycarbonate/polyetherimide, polyphenylene oxide, and polyphenylene oxide/polystyrene blends. The most prominent of the 1980s plastic material proposals were those with a General Electric origin, polycarbonate, polyetherimide, and polyphenylene oxide, all engineering plastics with good high temperature resistance. All eventually proved to be too expensive, even in combinations, for general food applications; and therefore after due testing and initial commercializations, all were removed from this application.

During this period of evaluating many different high temperature resistant plastics, crystallized polyester, developed originally during the 1970s for this very application, emerged as having the optimum combination of temperature resistance, chemical inertness, innocuous flavor, no measurable migration at any operating temperatures that could be of concern to regulatory agencies, and relatively good economics. Crystallization impairs some of the physical properties of the structure, e.g., impact resistance; and thus it is important to control the degree of crystallization, a temperature-limiting variable. This issue was addressed by coextruding crystallizable and amorphous polyester to generate a two-layer structure with good low-temperature toughness from the amorphous component. When the temperature is raised in the conventional oven, the amorphous polyester is partially crystallized to elevate its temperature resistance. Although this concept was conceived and implemented, its cost was too high for the benefit derived; and therefore it, too, was terminated. Crystallizing polyester slows the thermoforming process time, but this issue was readily overcome by increasing the size of the thermoformers.

The advantages of the crystallized polyester trays are sufficient for them to be one of the structures of choice for dual-ovenable, and now, the increasing microwave-only package structures. Polyester-coated paperboard and polyester film laminated molded pulp compete with crystallized polyester for this market, with polypropylene competing in the microwave-only sector of the market.

As tends to be the case, the temperature resistance of the various commercial alternatives is marginal, but acceptable; and potential exists for a higher temperature-resistant plastic structure which, of course, possesses all of the other attributes required for food packaging in the operating environments.

V. MICROWAVE SUSCEPTORS

Overcoming one of the major objections to microwave heating, a lack of surface browning or crisping when warranted, has been a true technical, legal, and marketing challenge. To achieve the desired effect requires that the temperature of the food surface in question be raised to at least 100°F above the boiling point of, water, with better flavor, aroma, mouth feel, and appearance resulting from the higher temperatures. Further, often the higher the temperatures, the shorter the time required for the effect. For example, in some radiation heaters generating temperatures of nearly 700°F at the source, desired surface browning of pizza may be effected in less than 40 s.

The first efforts to develop surface effects in microwave ovens occurred during the 1950s when ceramic materials characterized by high microwave energy loss were heated simultaneously with, and in contact with, the food. The food was heated by microwave energy and by conduction from the ceramic. The ceramic was also heated by exposure to the microwave energy. Because the upper temperature of the ceramic was not limited by water, the temperature could continue to rise to the 500 to 1000°F range where browning and surface crisping took place at very high speeds. This concept was translated into package materials but with little commercial success since, at that time, the market was nonexistent. Later, during the 1970s, carbon incorporated into a package material substrate was demonstrated to be a microwave absorber which could rise in temperature sufficiently to be an effective browning device.

During this same period, the discovery that very thin aluminum or other metallization on a substrate could absorb microwave energy and increase in temperature to browning levels led to the development of microwave susceptors. These are vacuum metallized plastic substrates with the aluminum levels very closely controlled so that the structure does not become a barrier to microwave energy or transparent to microwave energy. The thin coating absorbs the energy and increases its temperature to the 300°F+ range, usually sufficient to effect some surface evaporation plus some browning and crisping while permitting some of the microwave energy to pass through the material to heat the interior of the food.

Although the key to the functioning of the microwave susceptor is the vacuum metallization, of great importance is the substrate which must have a temperature limit of nearly 400°F as well as being transparent to microwave energy. The first of the requirements — temperature resistance — is met by oriented polyester film, which from past experience should also be inert, emit no undesirable odors, and therefore be safe. The reality, however, was that although polyester film had been evaluated thoroughly at ambient temperatures, few tests had been conducted at the high temperatures reached during susceptor browning. Although microwave susceptors had been on the commercial market for several years and had generated no particular consumer objections, during the late 1980s several technical professionals realized that none of the structures casually accepted for susceptors had been evaluated at the high temperatures; and, as a result, regulatory agencies began an intensive investigation into the chemistry of microwave susceptors. This investigation has been proceeding for more than 5 years with no untoward results, and the FDA has not elected to remove it from commerce because, to date, they have found nothing to imply any lack of safety, as they have so stated.

One objective that almost no one has meaningfully attacked is that almost all susceptors today require contact of the material with the product. This deficiency leads to nonuniform browning since not all of the package surface can be in intimate contact with an almost always irregular food surface. One means to overcome this defect is to develop packages that could emit infrared radiation at a distance so that the entire food surface would be uniformly browned. The plastic substrate would have to be even more temperature resistant than those currently being used. Susceptors which can deliver much higher temperatures than aluminum have been demonstrated; and these could become the desired infrared emitters, but, of course, only in conjunction with high temperature plastics meeting all of the other criteria for food packaging.

The events surrounding the development of microwave susceptors are highly instructive since the temperature resistance of the plastic substrate had been established, but the chemical breakdown products of polyester and the adhesive system bonding it to its paperboard substrate had not even been considered. As a consequence, the entire technology was at very high risk of being removed from the commercial market. During this time frame other plastic package material suppliers were proposing their own substrates as, for example, General Electric with its polyetherimide film, probably with no more information on the high temperature effects of this material or its adhesive than with polyester.

VI. CONCLUSION

The recital of high temperature polymers as they apply to food packaging could continue into discussions of other plastics such as fluoropolymers; liquid crystal blends with polyester; polybutylene naphthalate polyester; amorphous nylons such as MXD6; nylon 6,6; nylon 4,6; and so forth. Each has been proposed as a packaging plastic with high temperature properties of interest.

We recognize that high temperature to a polymer chemist or technologist is much higher than the high temperatures wanted by the food packager. Presently, the temperatures offered by plastics that meet all of the other food packager requirements are just barely enough. Higher temperature performance is desired, but not at any sacrifice of safety or damage to the contained product. Further, there are tight limits on compromises that can be tolerated on performance beyond safety.

The problems in the past have centered around ignoring or casually treating variables other than temperature performance. The plastics developer today and tomorrow will incorporate all variables into the development to ensure that the result can be acceptable to the food packaging community.

Chapter **11**

NEW HIGH-HEAT POLYCARBONATES: STRUCTURE, PROPERTIES, AND APPLICATIONS

Winfried G. Paul and Peter N. Bier

CONTENTS

I.   Introduction 203

II. Bisphenol TMC Polycarbonates 205
     A.   Synthesis 205
     B.   Structure 207
     C.   Properties 208

III. Applications 212

IV. Conclusion 218

References 218

I. INTRODUCTION

In 1953, Schnell[1] of Bayer AG and, independently, Fox[2] of General Electric Co. synthesized the first aromatic polycarbonates with attractive thermoplastic properties. Both had selected the aromatic diol bisphenol A (BPA) for their studies. BPA, made from acetone and phenol, was readily available, since it was being used as a raw material for epoxy resins. Marketed under the respective trademarks as Makrolon® resin and Lexan® resin, BPA polycarbonates have been tremendously successful. Their balance of key properties — excellent ductility, optical clarity, dimensional stability, heat resistance up to 148ºC, and reasonable price — made them engineering thermoplastics of choice for many purposes. Global polycarbonate consumption in 1993 approached 1.5 billion pounds,[3] with annual growth rates exceeding 10%, on the average, for the last two decades and amounting to more than twice the average growth of all plastics.

While BPA polycarbonates, with their attractive combination of desirable properties and economic production, still find new applications today, more than four decades after their conception, there have been requirements these resins could not fill. Market demand for transparent products of higher heat resistance stimulated intense research and development in the 1960s. This led to the market introduction of polysulfones (PSO), polyarylates (PAR), polyethersulfones (PES), and polyetherimide (PEI).

For polycarbonate producers, it seemed natural to modify the structure of polycarbonate, with the objective of maintaining the excellent properties, while gaining a significant increase in heat resistance. Until recently, plastics research at Bayer AG pursued this goal following essentially two alternative concepts:

- Reducing the conformational mobility around the carbonate function
- Increasing the persistence length

The former concept was applied in the development of tetramethyl bisphenol A polycarbonate.[4] The structural formula (1) demonstrates the steric hindrance of the carbonate group by methyl groups in all ortho positions. Accordingly, the glass transition temperature of the resin is raised to 203ºC. Unfortunately, ductility is greatly reduced.

$$(1)$$

In line with the latter concept, partial substitution of carbonate segments with aromatic dicarboxylic acid segments increases the rigidity of the polymer chain. Structure (2) indicates the respective persistence lengths (L_1, L_2) of both segments. Introducing aromatic ester segments raises the glass transition temperature; however, it also raises the melt viscosity, which makes these polyestercarbonates[5] (PEC) harder to process.

(2)

A third concept for modifying BPA polycarbonates to achieve higher heat resistance is based on:

• Restricting the conformational mobility of the phenyl rings

This route has recently been taken by Bayer AG and led to the invention of new polycarbonates, based on a new aromatic diol which is derived from 3,3,5-trimethylcyclohexane-1-one.[6]

II. BISPHENOL TMC POLYCARBONATES

A. Synthesis

The same raw materials from which BPA is derived, phenol and acetone, are also essential to the synthesis of the new bisphenol. Condensation of acetone (3) yields isophorone (4), which, in itself, is an important industrial solvent with an annual consumption (1993) of larger than 150 million pounds. Isophorone can be selectively hydrogenated to produce 3,3,5-trimethylcyclohexane-1-one, or TMC-one (5). This intermediate is then condensed with phenol to produce a derivative which is appropriately called bisphenol TMC (6).

Polycarbonates from bisphenol TMC (BPTMC) can be produced by the same methods and in the same production units which are suitable for BPA polycarbonates (PC).[7] A method of choice is an interfacial condensation process, in which phosgene is reacted with the sodium salts of the aromatic diols in the presence of tertiary amine catalysts. Cocondensation with BPA yields statistical copolycarbonates. The heat resistance, which

(3)   (4)   (5)

(5)   (6)

is a function of the glass transition temperature (T_g) of the resins, can be tailored by choosing the appropriate ratio of diols. See structural formula (7) of the copolycarbonates and Table 1 for the correlation of diol ratio and T_g.

(7)

TABLE 1.

Diol Ratio and Glass Transition
Temperature

n (mol%)	T_g (°C)	Designation
100	238	PC-HT100
55	205	PC-HT55
35	185	PC-HT35
10	160	PC-HT10
0	150	PC

In this chapter, the name for the new family of amorphous high-heat polycarbonates is abbreviated as PC-HT. The homopolycarbonate based on BPTMC has a glass transition temperature of 238°C. This resin can be referred to as PC-HT100. The copolycarbonates are designated accordingly (e.g., PC-HT35 stands for a resin which contains 35 mol% BPTMC and 65 mol% BPA). The products are marketed by Bayer and Miles under the trademark Apec® HT resins and are available in different grades for various requirements.

B. Structure

One of the outstanding features of bisphenol TMC polycarbonates is their low melt viscosity, which sets them apart from all known amorphous thermoplastics with higher glass transition temperatures than PC. Figure 1 shows melt viscosity at 360°C and a shear rate of 1000 s^{-1} for several high-heat thermoplastics, compared to compositions of the PC-HT family which have corresponding softening temperatures. The lowest melt viscosity is observed with PC-HT resins and can be described as a function of BPTMC content. This rheological behavior, in relation to the heat deformation resistance, has been explained by steric effects:[8] the three-dimensional structure formula (8) of the bisphenol TMC segment demonstrates 1,3-diaxial steric interaction between one of the cyclohexane methyl groups and one phenyl ring.

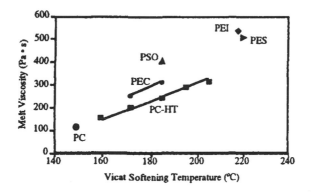

FIGURE 1
Melt viscosity at 360°C melt temperature and a constant shear rate of 1000 s^{-1} is plotted against the Vicat B softening temperatures of high-heat polymers. PC-HT resins have the lowest viscosity at a given softening temperature.

$$(8)$$

This is believed to impede the rotation of that phenylene ring about its 1,4-axis, whereas the equatorial phenylene is free to rotate. Also, the cyclohexane ring is indeed frozen in the chair conformation. Both effects account for the high T_g of PC-HT resins. The mobility of the carbonate

segments is equivalent to BPA polycarbonate, explaining the good ductility. It has also been stated that the three methyl substituents on the
cyclohexane ring might open the polymer matrix structure and increase
the fractional free volume (FFV) of the polymer.[9] In fact, the density of
PC-HT100 was found to be 1.107, compared to 1.200 for standard PC; and
the FFV was determined to be 16% larger than that of PC. In the melt, the
different restraints of molecular motion are relieved in steps, and flow
increases significantly with temperature. At 400°C, the melt viscosity is
independent of the BPTMC content, as can be seen from Figure 2.

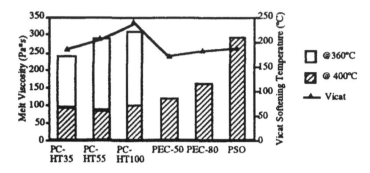

FIGURE 2.
For PC-HT resins, melt viscosity at a shear rate of 1000 s⁻¹ becomes independent of BPTMC
content at 400°C and is significantly lower than for other high-heat resins. The line graph
shows the corresponding Vicat B softening temperatures.

C. Properties

All PC-HT compositions are amorphous resins. Characteristic properties of BPTMC polycarbonates include:

- Excellent clarity
- Low intrinsic color
- Good thermal and UV-light stability
- Good dimensional stability
- Good ductility (also at low temperatures)
- Low density
- Low birefringence

In addition to the benefits of higher heat resistance, most of the property profile of bisphenol A polycarbonates has been maintained or even
further improved. With increasing BPTMC content, however, notch sensitivity is more prevalent than with standard PC. Ductility measured by
the multiaxial (instrumented) impact test shows comparable values for
PC and PC-HT resins. Table 2 summarizes how glass transition temperature, deflection temperature under load, notched Izod impact strength,

TABLE 2.

Physical Properties[a] of PC–HT and Other Amorphous Resins

	PC	PC-HT10	PC-HT22	PEC	PC-HT35	PSO	PC-HT55	PEI	PES
Glass transition (T_g), °C	150	160	172	179	185	189	205	219	222
DTUL @ 1.82 MPa, °C	132	140	150	162	162	174	179	190	195
Notched Izod @ 23°C, kJ/m²	45	30	14	25	8	5	7	4	6
Density, g/cm³	1.2	1.18	1.17	1.2	1.15	1.24	1.14	1.27	1.37
Refractive index	1.586	1.581	1.578	1.6	1.572		1.565		1.65

[a] These items are provided as general information only. They are approximate values and are not part of the product specification.

density, and refractive index vary as a function of BPTMC content. Other amorphous high-heat thermoplastics are included in the comparison.

Optical properties of bisphenol TMC polycarbonates compete very favorably with those of other transparent resins. Figure 3 compares yellowness indices and total light transmission values with polymethacryl methylimide (PMMI), PSO, and PEC-80 (80% aromatic ester segments), and PEI — as molded, after 2000 h of oven aging at 150°C, and after 2,000 h exposure to artificial weathering. The superior UV stability of PMMI is marred by its embrittlement in the form of cracking.

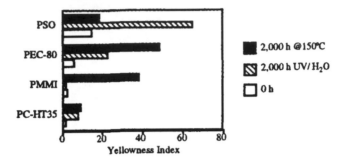

FIGURE 3
PC-HT35 shows the least yellowing in a comparison of heat aging and artificial weathering results for transparent high-heat resins.

Another optical property, birefringence, is caused by both polymer chain orientation and stress in the sample resulting in anisotropic electron polarization. Consequently, this anisotropy can be separated into a rheo-optical contribution (ROC), which is related to molded-in stress, and a material contribution. The rheooptical constant of PC-HT copolycarbonates as a function of BPTMC content is shown in Figure 4. It characterizes the ease of processing for low birefringence due to molecular orientation.

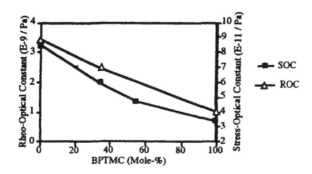

FIGURE 4
The rheo-optical constant (ROC) for PC-HT resins decreases with increasing BPTMC content. The same is true for the stress-optical constant (SOC).

Figure 4 also shows the dependency of the stress-optical constant (SOC), which characterizes the electron polarization of the molecule itself. In both cases, values for PC-HT resins are significantly lower than for PC.

A summarizing comparison of key properties of PC-HT resins with PEC, PAR, PSO, PES, and PEI shows the characteristic strengths and weaknesses of each class. Although details had to be neglected, Table 3 differentiates between the polymers for preselection purposes. It turns out that PC-HT resins are superior in all regards except flame and chemical resistance.

TABLE 3.

Summary of Properties for Amorphous High-Heat Polymers

	PC-HT	PEC/PAR	PSO/PES/PEI
Processibility	+	−	−
Optical clarity	+	−	−
UV stability	+	−	−
Price	+	+	−
Ductility	+	+	−
Flame resistance	−	−	+
Chemical resistance	−	−	+

Unexpectedly, the amorphous PC-HT products also show certain advantages in dimensional stability over crystalline, reinforced thermoplastics. At room temperature, amorphous resins generally exhibit lower rigidity and higher coefficients of expansion than semicrystalline resins, especially when those are filled or reinforced. It is evident from the storage modulus curve in Figure 5 that this is not the case at a service temperature of 150°C: unfilled PC-HT35 has a higher modulus than a polyphenylene sulfide (PPS) filled with 55% mineral and glass. As a related property, the coefficient of linear thermal expansion (CLTE) of amorphous PC-HT stays virtually constant from room temperature to softening temperature, whereas the CLTE of semicrystalline resins changes due to glass transitions below the crystalline melting point.

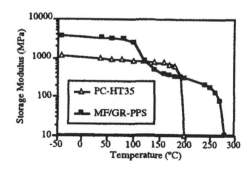

FIGURE 5

The dynamic-mechanical analysis of PC-HT35 and mineral/glass-filled PPS reveals the higher storage modulus of the amorphous resin in the range from 140°C to 180°C.

PC-HT resins are compatible with many other thermoplastics, including standard PC, PBT, PET, ABS, and their blends. This feature contributes to recyclability, conceptually allowing for recycling of modules,[10] which are composed of different compatible resins. Compared to standard polycarbonates, PC-HT resins also exhibit higher compatibility with many polyolefins, such as polyethylene, polypropylene, polyisobutylene, as well as ethylene-propylene-monomer (EPM) rubber and ethylene-propylene-difunctional monomer (EPDM) rubber. This change in solubility characteristics can be explained by the more aliphatic nature of BPTMC and its compounds. Table 4 shows that solubility of the PC-HT resins in many organic solvents is also improved over BPA polycarbonate and increases with BPTMC content.

TABLE 4.

Solubility of PC-HT Resins in Organic Solvents

Solvents	PC	PC-HT35	PC-HT55	PC-HT100
Carbon disulfide	–	–	0	N/A
Carbon disulfide/acetone	–	+	+	N/A
Dimethylformamide	–	+	+	N/A
Ethyl acetate	–	+	+	+
Methyl ethyl ketone	–	+	+	+
Methylene chloride	+	+	+	+
Tetrahydrofuran	+	+	+	+
Toluene	–	+	+	+

III. APPLICATIONS

The development of PC-HT resins originated from the demand for higher heat resistance in applications, which extended beyond the limits of standard PC. PC-HT applications can be classified by the duration of heat exposure and can be roughly and somewhat arbitrarily divided into the following service categories:

- Long-term heat
- Mid-term heat
- Short-term heat
- No heat (PC-HT selected for other reason)

Depending on the category required, a PC-HT grade will have different upper service temperature limits. For long-term exposure, relative temperature indices, such as Underwriters' Laboratories RTI per UL 746B or the temperature index (TI) of the International Electrotechnical Commission per IEC 216, can be used as a guide (Table 5).

TABLE 5.

Temperature Indices

Standard	Property Tested	PC	PC-HT10	PC-HT35
UL 746 B RTI	Electrical	125	140	150
(40,000 h)	Mechanical with impact	115	130	130
	Mechanical without impact	125	140	150
DIN IEC 216 TI	Electrical	135	—	150
(20,000 h)	Mechanical with impact	110	—	140
	Mechanical without impact	125	—	150
DIN IEC 216 TI	Electrical	145	—	160
(5,000 h)	Mechanical with impact	125	—	150
	Mechanical without impact	145	—	160

Table 5 also shows the time span for the property under test to reach 50% of the original value, which is the criterion used to determine the temperature index. Temperature indices for several grades based on PC-HT10 and PC-HT35 have already been listed. UL and IEC standards are criteria used especially for electrical, electronic, and lighting applications, such as connectors and enclosures, e.g., reflector housings for plant lighting or recessed residential lighting. A service temperature of 140ºC and additional requirements for flame-retardant behavior were the demands for the housing of a halogen bulb-recessed lighting fixture (Figure 6). An opaque flame-retarding grade of PC-HT35 fulfilled this requirement with a UL 94 V-0* rating at 1.6 mm and also passed the glow-wire ignition test per IEC 695-2-1. (A transparent grade with UL 94 V-0* rating at 3.2 mm is also available.)

Consumer applications for small appliances are typically specified for mid-term exposure; and, in the case of hair dryer and steam iron parts, also have to pass short-term torture tests at peak temperatures. For the latter, deflection temperatures under load (see Table 2) can give guidance, although final grade selection has to be done by testing of the actual application part. Further requirements for a hair dryer diffusor (Figure 7) included processibility in tools designed for standard PC, dimensional stability, surface gloss, and the ability to color-match parts with others made from PC. A very specific added benefit of selecting PC and PC-HT was the possibility to sandblast molded parts to achieve a gunmetal-like surface for high-end hairdryer models.

---

* Flammability results are based on small-scale laboratory tests for purposes of relative comparison and are not intended to reflect the hazards presented by this or any other material under actual fire conditions.

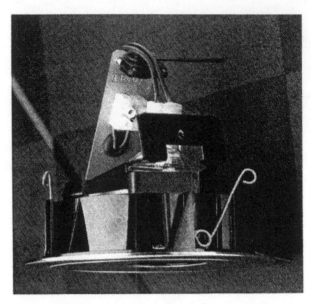

FIGURE 6
A recessed lighting fixture for halogen bulbs produced by Lumiance (NL) uses a housing made from a flame retarding grade of PC-HT35.

FIGURE 7
Braun AG (FRG) selected PC-HT35 for the diffusor of this hair dryer.

Automotive lighting parts are also targeted for mid-term service life, with applications including lenses and reflectors. The standard thermoplastic lens materials in this field, polymethyl methacrylate and polycarbonate, very often cannot stand up to the peak temperatures generated by newly designed aerodynamic headlamps or stylish taillamp assemblies, given the growing demands on light output. The optical properties of automotive lens materials must meet the stringent requirements of the transport authorities, such as the Federal Motor Vehicle Safety Standard (FMVSS) in the U.S., or the European Community (ECE) standards.

For outer lens applications in forward lighting, PC-HT resins, just like standard PC resins, must be additionally protected with hard (scratch-resistant) coatings. Several grade/coating combinations have already passed the requirements of the ECE Regulation 19-02 and are being weathered in Florida and Arizona to pass the FMVSS-required exposure tests per standard Society of Automotive Engineers (SAE) J576c. An example of PC-HT10 in an inner lens of a taillight assembly can be seen in Figure 8. Here the prisms which create the light pattern have been moved to an inner lens to allow a smooth outer appearance. The design forms a sealed twin-wall cover not unlike a thermopane window, with the same consequence of increasing the temperature on the inside. For all lens applications, heat resistance and UV stability are essential. While the hard coats provide UV protection from sunlight on top of basic UV stability (which is enhanced by state-of-the-art UV-absorbing additives), additional UV exposure from halogen bulbs should be reduced, since the combination of heat and UV is synergistic in its detrimental effect. So-called UV cutoff bulbs are available for this purpose.

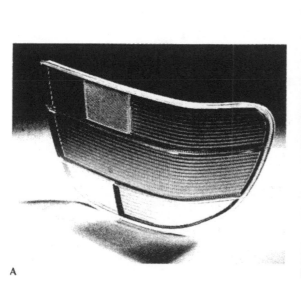

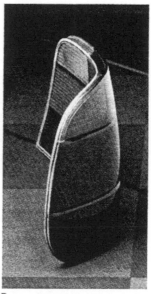

A

B

FIGURE 8
The new luxury-class BMW models have taillights assembled from an acrylic outer lens and a prismatic inner lens molded from PC-HT10.

Reflector housings typically demand good dimensional and vibration stability, excellent surface quality, and preferably the ability to be metallized without a base coating. Further demands include low density to reduce part weight at the front end of the car, which is essential for energy conservation.

Applications in medical and related markets, with requirements to sterilize parts by various methods, are another field for PC-HT resins. While standard PC parts are often subjected to one-time steam sterilization* at 121ºC, higher steam autoclave temperatures (134ºC, 143ºC) lead to deformation. Here PC-HT22 is being used with the advantage of shorter autoclave cycles, e.g., in animal breeding cages. Medical grades of PC-HT resins meet the United States Pharmacopeia (USP) XXIII, Class VI test requirements for biocompatibility, as well as the more extensive Tripartite testing. Several housing parts of the laparoscope in Figure 9 have also been molded from PC-HT22. As a general caution it should be added that in steam sterilization, as in prolonged water contact at temperatures above 65ºC, the use of alkaline and amine additives has to be avoided to prevent hydrolysis of PC and PC-HT.

An added advantage of PC-HT resins is their stability against γ-irradiation, which is an alternative method for sterilizing parts. Special tints are offered for this purpose.

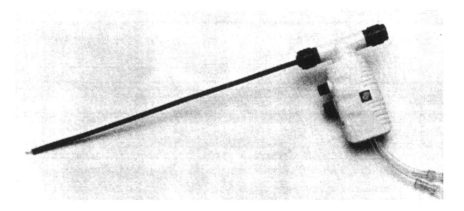

FIGURE 9
The housing of Birtcher Medical Systems laparoscope is made from PC-HT22 to allow steam sterilization at 134ºC.

* The sterilization method and the number of sterilization cycles a part made from a resin can withstand will vary depending upon type/grade of resin, part design, and processing parameters. Therefore, the manufacturer must evaluate each application to determine the sterilization method and the number of sterilization cycles appropriate for exact end-use requirements.

PC-HT formulations which have been rendered conductive or antistatic by incorporating metal or carbon fibers, are being used in medical applications which are also suitable for electron beam sterilization. Another application of carbon fiber blends is in the field of electronics packaging, where the curing of integrated circuits requires high dimensional stability and antistatic properties for the so-called chip carrier trays.

A short-term peak heat requirement during the production process characterizes the following applications. PC-HT resins are being evaluated as surface-mounted LED (light-emitting diode) housings, which are designed to withstand short-term exposure to vapor phase or IR soldering conditions (above 215°C) in the production of circuit boards. An extraordinary advantage of higher heat resistance, in addition to typical PC properties, has been uncovered in recent work on the substitution of glass, e.g., for automotive glazing. Raising the cure temperature of hard coats shortens the cure time and/or dramatically increases the surface hardness of the coatings. Special formulations actually approach the abrasion resistance of glass in the Taber test. Figure 10 shows the effect of increased cure temperatures in a comparison of abrasion tests performed on two different coatings.

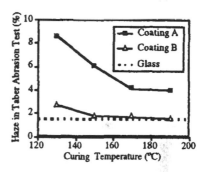

FIGURE 10
The abrasion resistance in the Taber test is a function of the curing conditions with which hard coats are applied. A standard coating improves significantly with an increase in curing temperature, and special coatings reach the scratch resistance of glass (1.5 % haze).

The solubility of the new resins has only recently raised interest. Advantages of PC-HT resins include the ability to use non-halogenic solvents to reduce hazards to workers and environment, as well as a reduced crystallization tendency. Special grades of high purity and higher molecular weight have been made available for solution applications. Examples include use as binder resins for organic photoconductors, e.g., in laser copier drums, or for other heat-conducting and electrical current-conducting layers. Evaluations for use in membranes have revealed high permeability for various gases (N_2, O_2, He, CH_4, CO_2) compared to most polycarbonates. Fairly low permselectivity has been measured due to the "open structure" of PC-HT100.

The reduced birefringence of BPTMC polycarbonates is likely to be applied in the field of magnetooptical recordings, where today glass or low-molecular-weight standard PC is in use. To demonstrate the advantage of PC-HT35 over standard PC, compact discs were molded from both resins. Measuring the optical path difference over the radius at a service temperature of 100°C shows that PC-HT35 does not exceed a deviation of ±1 nm, which will be required for data storage discs in the future (see Figure 11).

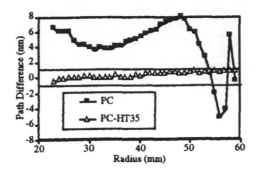

FIGURE 11
The optical path difference, as measured over the radius of a compact disc at 100°C, is significantly lower for PC-HT35 than for PC

IV. CONCLUSION

A novel structure has been introduced into the standard polycarbonate chain, which extends the service range of this polymer class to significantly higher temperatures and new applications. Technological and economic benefits of this development have just begun to appear, and exciting possibilities are on the horizon.

REFERENCES

1. Schnell, H., Bottenbruch, L., and Krimm, H., German Patent, 971,790, 1959.
2. Fox, D. W., History and commercial development of polycarbonates, presented at SPE/ANTEC, Chicago, IL, 1985.
3. Resins 1994: Plotting a course for supply, *Mod. Plast.*, 1, 45, 1994.
4. Serini, V., Freitag, D., and Vernaleken, H., *Angew. Makromol. Chem.*, 55, 175, 1976.
5. Markezich, R. L. and Quinn, C. B., German Patent, 3,016,020, 1979.
6. Freitag, D. and Westeppe, U., A new principle for polycarbonates with superior heat resistance, *Macromol. Chem. Rapid. Commun.*, 12, 95, 1991. Kämpf, G., Freitag, D., and Fengler, G., New polycarbonates with improved heat resistance, *Kunsts. Plast. (Munich)*, 5, 1992.
7. Freitag, D., Westeppe, U., Wulff, C., Fritsch, K.-H., and Casser, C., U.S. Patent, 4,982,014, 1991; U.S. Patent, 5,126,428, 1992.

8. Freitag, D., Fengler, G., and Morbitzer, L., Routes to new aromatic polycarbonates with special material properties, *Angew. Chem. Int. Ed. Engl.*, 30, 1598, 1991.
9. Hägg, M.-B., Koros, W. J., and Schmidhauser, J. C., Gas sorption and transport properties of bisphenol-I-polycarbonate, *J. Polym. Sci., Part B*, 32, 1625, 1994.
10. Paul, W. G. and Bier, P. N., Amorphous high-heat polycarbonates for advanced lighting applications, *Proc. SPE/ANTEC*, 52, 1759, 1994.

Chapter **12**

WEAR PERFORMANCE OF HIGH TEMPERATURE POLYMERS AND THEIR COMPOSITES

Klaus Friedrich

CONTENTS

Abstract 222

I. Introduction 222

II. Wear Testing Procedures 223

III. Sliding of High Temperature Polymers vs. Steel..................223
   A. Effects of Filler and Matrix Related Variables on
      Friction and Wear at Room Temperature......................223
      1. Polymeric Matrices 223
      2. Lubricants and Reinforcements 225
      3. Fretting Wear of LCP Matrix Systems for
         High-Precision Bearings 228
   B. Polymeric Materials for Tribological Use at
      Elevated Temperatures 236
      1. Polyimide, Poly(Ether Ketone), and
         Polybenzimidazole..236
      2. Design Example for Sliding Shoe........................236
   C. Continuous Fiber/Polymer Matrix Composites 239
      1. Systematic Development of Hybrid Composites 239
      2. Filament Would Bearings of High Load Bearing
         Capacity...241

IV. Conclusions 243

Acknowledgments 244

References 244

ABSTRACT

Polymeric materials containing different fillers and/or reinforcements
are frequently used for applications in which friction and wear are critical
issues. This overview describes especially how to design high temperature resistant thermoplastics, e.g., filled with carbon fibers and internal
lubricants, in order to operate under low friction and wear at elevated
temperatures as sliding elements in, e.g., textile drying machines. Further
information will be given on the sliding wear behavior of some liquid
crystal polymer (LCP) systems against steel, and on special studies of
fretting wear behavior of various LCP composites, as used for high
precision bearing systems, e.g., in stepper motors or cooling fans. Finally,
a systematic development of continuous fiber/polymer composites, to
be used as filament would bushings (corrosion resistant, maintenance
free) or as spherical plain bearings of extremely high load carrying capacity (thin layer composites) is outlined.

I. INTRODUCTION

Tribology is the science that deals with the design, friction, wear, and
lubrication of interacting surfaces in relative motion (as in bearing or
gears). Composite materials, one of the most rapidly growing classes of
materials, are being used increasingly for such tribological applications.
Yet, by now, much of the knowledge on the tribological behavior of composite materials is empirical, and very limited predictive capability currently exists. Nevertheless, in recent years, it has been attempted to determine to what degree phenomena governing the tribological
performance of composites can be generalized and to consolidate inter-
disciplinary information for polymer-, metal-, and ceramic matrix composites.[1] The importance of promoting better knowledge in the mundane
areas of friction, lubrication and wear, in general, has been further demonstrated by quite a number of important articles and books published
on these topics in recent years (see, e.g., References 2–8). Along with the
comprehensive *Encyclopedia of Composite Materials*[9] and a variety of other
tribology handbooks cited in a book entitled *Friction and Wear of Polymer
Composites*,[10] these references should build a powerful basis for finally
understanding the complexity of composites tribology in the same way
as is the case for isotropic materials.

The topic of primary concern here is how various structures and
compositions of polymer composites influence their friction and wear

behavior. In particular, the discussion will be confined to dry sliding of polymer-based materials against smooth steel since these systems have received more attention, and developed further, during the past few decades than any other group.

II. WEAR TESTING PROCEDURES

The choice of the type of wear test configuration must be based on the tribotechnical system under consideration (Figure 1). The latter determines the elements of the basic structure of the tribosystem which yields information on the existing wear mechanisms (surface variations), and the loss of material (wear rates). In most of the studies reported here three types of test configurations were chosen for material evaluation: (1) pin-on disc, (2) block on ring, and (3) block on block. The first two were used for situations in which the polymeric component was slid on one and the same track against a steel counterpart. Procedure (3) refers to the condition in which a polymeric component in oscillation moved (over a short amplitude) against the metallic mating material. In both cases, the tests can be conducted under controlled temperature conditions, as described in References 11 and 12. Most often, the specific wear rate $\dot{w}_s$ of the material to be optimized is calculated by the equation

$$\dot{w}_s = \frac{\Delta v}{F_N \cdot L}[mm^3/(N \cdot m)]$$

where

        ΔV = loss in volume
        F_N = normal load
        L   = sliding distance

The inverse of the wear rate is usually referred to as the wear resistance of a material.

III. SLIDING OF HIGH TEMPERATURE POLYMERS VS. STEEL

A. Effects of Filler and Matrix Related Variables on Friction and Wear at Room Temperature

1. Polymeric Matrices

Besides the traditional polymeric materials used as matrices for friction and wear-loaded components (e.g., polyacetale [POM], polyamide [PA],

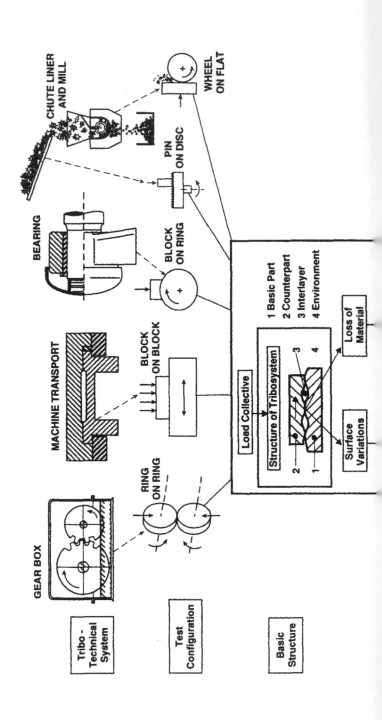

poly(tetrafluoroethylene) [PTFE][13-17]), now newer high-performance polymers have found more and more entrance into special tribological applications in which high service temperature is a critical issue (e.g., polyetherimide [PEI], polybenzimidazole [PBI], liquid crystal polymers [LCP], and polyimide [PI][18-21]). Very special candidates in this respect are also the poly(arylether ketones) which allow tailoring of their glass transition and melting range by variation of ether and ketone groups in the polymer chain (Figure 2).[22] Figure 3 relates the friction and wear data of PEEK and PEEKK to those of other high temperature resistant polymers. Further effects on the wear resistance of the neat PEEK itself can arise from the molecular weight of the material and other morphological parameters, such as degree of crystallinity and preorientation.[23] For the same degree of crystallinity, an increase in molecular weight (that corresponds to an increase in melt viscosity) yields a reduction of depth wear rate under various conditions of contact pressure p and sliding speed v (Figure 4).[24]

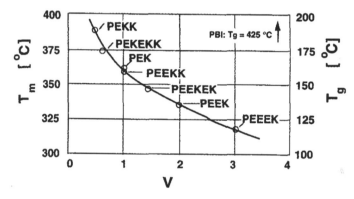

FIGURE 2
Melting and glass transition temperature of various poly(arylether ketones) as a function of ether/ketone ratio. (From Münstedt, H. and Zeiner, H., *Kunststoffe*, 79, 993, 1989. With permission.)

2. Lubricants and Reinforcements

The addition of short fibers to these high-temperature and wear-resistant polymers is mainly intended to improve properties other than wear, such as stiffness, impact resistance, thermal conductivity, and creep resistance at elevated temperatures.[25] Additions of solid lubricants, in particular graphite, PTFE, and MoS_2, for reduction of the friction coefficient usually degrade the mechanical properties but this can be offset by fibrous reinforcement. A vast range of polymer/solid lubricant/fiber composites in now available, presenting the designer with a confusing choice since the properties of many of these composites tend to be very similar.[26] A few examples of how PTFE affects the friction and wear behavior of

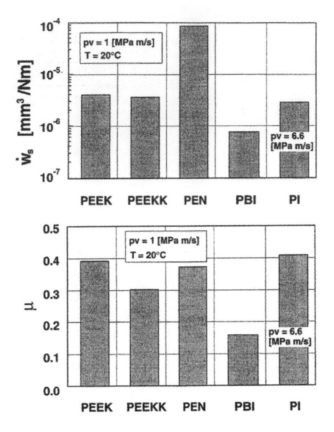

FIGURE 3
Comparison of specific wear rates and coefficients of friction of various high-temperature polymers, tested at room temperature, a pressure of p - 1 MPa and a sliding speed of v = 1 m/sec. (From Lu, Z., Institut für Verbundwerkstoffe, Universität Kaiserslautern, Germany, 1994. With permission.)

engineering polymers by transfer film formation are given in Figures 5 to 7.[27] It illustrates that the ultimate steady-state rate of wear is only partly related to its bulk structure but depends also on the surface conditions generated by the sliding process itself.

The effects of various short fiber reinforcements on the sliding wear characteristics of these polymers are shown in Figures 8 and 9. In those polymers that possess high specific wear rates in the unreinforced condition almost any type of reinforcing fiber results in significant reductions in wear and improvement to mechanical properties. In addition, carbon fibers often improve the material's wear resistance by a factor of five or more than the same volume fraction of glass fibers.[28] From recent studies carried óut with different types of carbon fibers (PAN based vs. Idemitsu HT pitch-based CF) it can be concluded that the composite wear resistance will be better, the less sensitive the fibers react against fiber/fiber friction and early breakage. In addition, a strong bonding to the matrix helps to

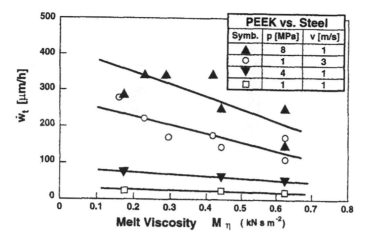

FIGURE 4
Reduction of depth wear rate of PEEK at room temperature due to an increase in molecular weight. (From Lu, Z., Institut für Verbundwerkstoffe, Universität Kaiserslautern, Germany, 1994. With permission.)

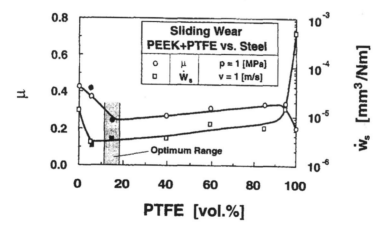

FIGURE 5
Influence of PTFE filler content on specific wear rate and coefficient of friction of PEEK. (From Lu, Z., Institut für Verbundwerkstoffe, Universität Kaiserslautern, Germany, 1994. With permission.)

maintain broken fiber pieces in the composite surface, thus preventing early formation of third body abrasives and enhanced wear (Figures 10 and 11).[29,30] Similar effects of glass fiber reinforcement on wear of injection-molded thermoplastics have also been found for LCPs. In this case, the high anisotropy of the LCP matrix is also detectable with regard to its specific wear rate (Figures 12 and 13). In addition, it was found by Luise[31] that LCPs blended with tetrafluoroethylene polymers possess improved wear properties.

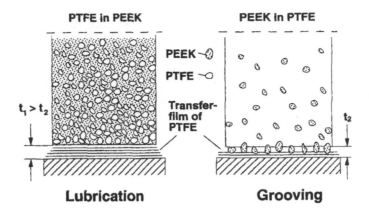

FIGURE 6
Schematic illustration of transfer film formation of different thickness (t) in case of low PTFE content in PEEK (t_1) vs. PTFE matrix with PEEK inclusions ($t_2 < t_1$). (From Lu, Z., Institut für Verbundwerkstoffe, Universität Kaiserslautern, Germany, 1994. With permission.)

3. Fretting Wear of LCP Matrix Systems for High-Precision Bearings

Besides unidirectional sliding, bidirectional sliding with small slip amplitudes, i.e., fretting, is of interest for the application of LCPs for high-precision bearings, e.g., in stepper motors. The fretting wear behavior of neat and reinforced LCPs (Table 1) has been studied with motion parallel and transverse (L and T orientation) to the mold-filling direction on the injection-molded surface. The tests were performed with a sphere on flat configuration, and the wear was characterized by the penetration depth of the sphere. To characterize the fretting wear behavior as close as possible to the real in-service conditions of the aimed bearings, the test parameters were adjusted to: diameter of the sphere d = 10 mm, normal load F = 50 N, frequency f = 20 Hz, slip amplitude A = 350 µm, test duration t = 4 h, and environmental temperature T = RT.

Depending on the kind of LCP, the resulting fretting friction and wear behavior are different. Highest wear resistance is found for the base polymer of the VECTRA series (LCP-A). Compared to the other LCP materials, the specific wear rate elaborated for this material is lower by half an order of magnitude and the friction coefficient is lowest. Although the LCP exhibits a high molecular orientation for the injection-molded surface,[32-36] the fretting results do not show clear tendencies or remarkable differences between L and T direction. Therefore only the results of the L direction are presented here. The wear occurring is a result of the high strength within and the poor bonding between the fibrils which build up the skin layers of the LCP.[34,37] Under load the fibrils are peeled out of the surface. Due to the high strength of the fibrils, the worn material is not totally cut from the bulk. Bulges of worn material are built up on both return lines of the sphere. Profiles of the worn area obtained by laser profilometry show that the volume of the bulges can be in the same range as the volume

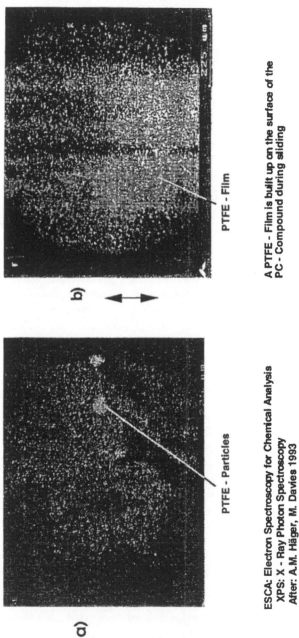

PTFE - Particles

PTFE - Film

A PTFE - Film is built up on the surface of the PC - Compound during sliding

ESCA: Electron Spectroscopy for Chemical Analysis
XPS: X - Ray Photon Spectroscopy
After: A.M. Häger, M. Davies 1993

FIGURE 7

Direct evidence of PTFE-transfer film formation in case of a PC-PTFE-blend slid against a hard steel surface. (From Häger, A. M. and Davies, M., *Advances in Composite Tribology,* Elsevier Science Publ., Amsterdam, 1993, 107. With permission.)

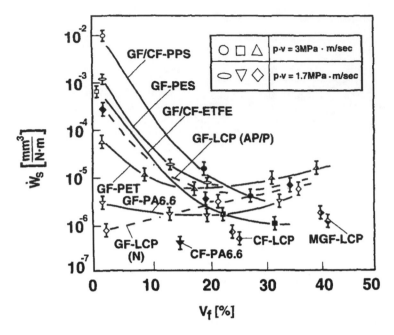

FIGURE 8
Specific wear rate as a function of short glass fiber content in various high temperature thermoplastics vs. steel. (From Friedrich, K., Ed., *Composite Material Series*, Vol. 1, Elsevier, Amsterdam, 1986. With permission.)

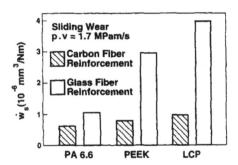

FIGURE 9
Comparison of the effect of glass relative to carbon fiber reinforcements on the specific wear rate of three different thermoplastic matrices. (From Voss, H., *Fort Schrittsberichte VDI*, Reihe 5: Grund-und Werkstoffe, Nr. 116, VDI Verlag, Düsseldorf, Germany, 1987. With permission.)

of the wear tracks (Figure 14). For the intended application as bearing material, this wear behavior is poor. Under in-service conditions the peeling of one fibril out of the surface will probably initiate further peeling of other fibrils. Additionally, the peeling effect will increase with increasing temperature due to the decrease in intermolecular bonding. The effect of environmental temperature has been studied up to 125°C. The specific wear rate increases by more than one order of magnitude in the studied

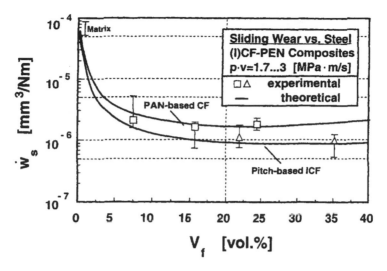

FIGURE 10
Specific wear rate of PEN composites, reinforced with two different types of carbon fibers. The error bars illustrate the experimental data, the fully drawn lines the theoretical predictions according to Reference 29. (From Friedrich, K., Karger-Kocsis, J., Sugioka, T., and Yoshida, M., *Wear*, 158, 157, 1992. With permission.)

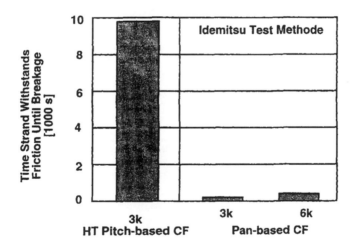

FIGURE 11
Time to failure under fiber/fiber friction testing of the two different types of carbon fibers used as reinforcements in Figure 10. (From Friedrich, K., as based on personal communication with Idemitsu Kosan Co., Tokyo, Japan, 1991. With permission.)

range (Figure 15). On the other hand, the friction coefficient decreases to below $\mu \approx 0.1$. The latter is an additional hint for the increased peeling of the fibrils. The fibrils can act here like rollers.

It is well known that the fibrillar structure of the skin layers is a result of the specific melt flow conditions during molding; and this fibrillation

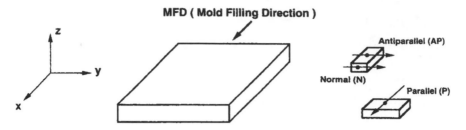

FIGURE 12
Various sliding directions on small samples, prepared from injection-molded plates.

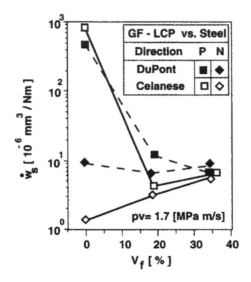

FIGURE 13
Sliding wear data of two various LCP systems, filled with short glass fibers. Sliding against steel with the major molecular axis normal to the sliding surface gives superior results compared to sliding on the plate surfaces with molecules in plane and parallel to the sliding direction. (From Voss, H. and Friedrich, K., *J. Mater. Sci.*, 21, 2889, 1986. With permission.)

can be suppressed using fillers or reinforcements.[38] Glass fibers as reinforcement are often used to increase the wear resistance. In case of an LCP matrix, glass fibers do not bring advantages in the friction and wear behavior in the aimed range (Figure 16). For small amounts of glass fiber reinforcement the wear resistance is just unaffected or at least slightly reduced. Further increasing contents of glass fibers in the matrix results in a small improvement in wear resistance, but, e.g., in case of the GF/LCP-A, the wear resistance of the neat polymer is not reached. The reduction of wear with increasing fiber content is a result of the specific wear condition. Under oscillatory motion the transport of debris out of the wear region is hindered; in addition, normally the low fiber matrix

TABLE 1.

LCP Material Compositions and Codes

Composition	Code
Neat Vectra® LCP (LCP-A)	Vectra A950
Neat Rhodester CL® LCP (LCP-R)	Rhodester CL
Vectra® LCP and ? wt% short glass fibers	? wt% GF[a]
Vectra® LCP and 25 wt% poly(tetrafluoroethylene)	25 wt% PTFE
Vectra® LCP and 30 wt% short carbon fibers	30 wt% CF
Vectra® LCP and 30 wt% talc	30 wt% Talc
Vectra® LCP and 20 wt% poly(tetrafluoroethylene) and 20 wt% short carbon fibers	20 wt% PTFE + 20 wt% CF
Vectra® LCP and 25 wt% poly(tetrafluoroethylene) and 15 wt% short glass fibers	25 wt% PTFE + 15 wt% GF

[a] Various filler contents.

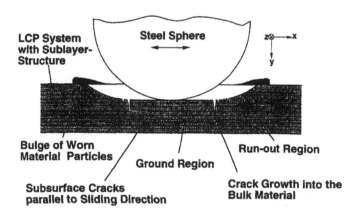

FIGURE 14

Schematic of wear mechanisms during fretting of an LCP surface by a steel ball. (From Schledjewski, R. and Friedrich, K., Proceedings of 13th International European Chapter Conference of the SAMPE, Hamburg, Germany, May 11–13, 1992, 93. With permission.)

adhesion results in easy damage of the glass fibers. Low amounts of glass fibers result in abrasive debris in the contact zone which can plough through the surface and increase the wear. Further increase in fiber content leads to an increase in debris amount; on the other hand, the abrasive particles cannot plough long distances through the surface without being stopped by the reinforcing fibers. A more or less pronounced decrease in wear can be found. As a result of the mechanisms explained, the friction coefficient exhibits also a slight maximum peak for low-fiber contents due to the long distance ploughing effect, especially for those materials which exhibit a pronounced maximum in specific wear rate.

Using other typical fillers or reinforcements for polymers, it can be seen that the fretting friction and wear properties of LCP-A get worse in most cases (Figure 17). Only the friction behavior can be improved by using PTFE as filler. Carbon fibers, which are often used to improve the

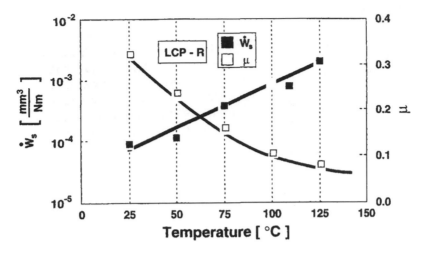

FIGURE 15
Effect of testing temperature on specific wear rate and coefficient of friction on LCP-R.

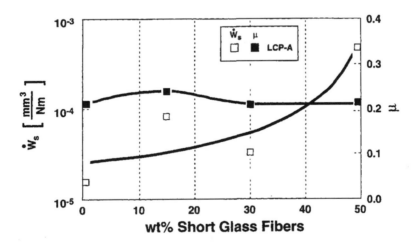

FIGURE 16
Influence of short glass fiber reinforcement on specific wear rate and coefficient of friction of LCP vs. steel at RT.

wear resistance, result in wear rates higher by nearly one order of magnitude. The typical filler material talc increases the wear additionally by one order of magnitude. While the behavior of the latter filler was expected (talc is the weakest mineral material known and fails very easily under shear load), the effect of carbon fiber reinforcement was unexpected. Similarly as it was found for the glass fiber reinforcement, the main problem is the poor bonding between fiber and matrix. Since the wear tests were performed on the injection-molded surface, the reinforcing fibers are aligned parallel to the wear surface. It seems that the fiber matrix adhesion is much worse than in the case of glass fiber reinforcement. It can be concluded that the load transfer into the poorly bonded fibers (glass as well as carbon) results in easy pullout of them and therefore in an increase in wear compared to the neat polymer. Using lubricants, the load transfers into the counterpart and therefore the wear can be reduced.[39] Besides liquid lubricants, very often solid lubricants like PTFE are used because in many applications a contamination of the environment by oil or grease has to be prevented. Furthermore, incorporated solid lubricants can be considered as a lifetime lubrication. In Figure 18 the effect of the combined use of fiber reinforcement and PTFE filling can be seen. In the case of glass fiber reinforcement, the additional PTFE filling exhibits nearly no effect. This can be explained by the abrasive character of broken glass fiber particles. The PTFE lubrication cannot prevent the failure of single fibers in the wear region. As soon as a single fiber has failed, the broken glass particles will act abrasively and initiate further glass fiber failure. In the case of carbon fiber reinforcement the failure of a single fiber will bring no problems. This fiber will be transported out of the wear region since it will not act in an abrasive way.

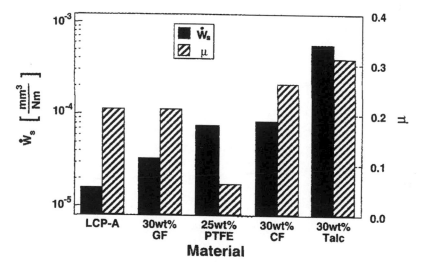

FIGURE 17
Effect of various types of fillers on specific wear rate and coefficient of friction of LCP-A vs. steel.

Concerning wear mechanisms, besides the build up of bulges of worn material, another effect typical for the fatigue load of the surface can be found (Figure 14). On both sides of the ground region of the wear track, cracks are initiated. Starting from these transverse cracks, subsurface cracks can be found which led to a kind of delamination of individual layers of the bulk. Using fiber reinforcement, the initiation of these transverse cracks can be reduced, but only in the case of PTFE filling does the

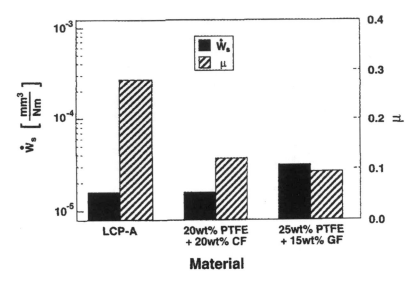

FIGURE 18
Comparison of neat LCP with PTFE/CF- and PTFE-/CF and PTFE-/GF-filled LCP-A.

reduction in transferred tangential force into the LCP surface suppress any initiation of the transverse cracks.

B. Polymeric Materials for Tribological Use at Elevated Temperatures

1. Polyimide, Poly(Ether Ketone), and Polybenzimidazole

A summary about how short fibers and internal lubricants can affect the specific wear rate and the coefficient of friction at room temperature and at 150°C testing temperature is given in Figure 19.[24] In addition, Figure 20 illustrates what high load bearing capacities can be achieved with some of these systems, even at 220°C test conditions. It seems that polymers with a higher glass transition temperature (T_g) (PBI > PEEKK > PEEK; see Figure 2) exhibit also higher resistance against sliding wear at elevated temperatures. A very potential candidate in this respect is PBI, although it is hard to process and performs quite brittle. Both problems reduce its applicability for parts with complex geometry that need to be used under conditions where impact resistance, besides tribological performance, is of a major concern.[40]

2. Design Example for Sliding Shoe

A particular application for which a self-lubricated polymer-based composite was considered to be a good alternative were sliding shoes in textile drying machines. The function of these shoes was to guide two heavy, 20 m long steel chains in a parallel motion through a drying oven that was used to dry wet printed textiles in a temperature range between

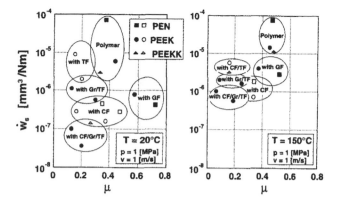

FIGURE 19
Summary of the effects of fiber reinforcements and internal lubricants on the coefficient of friction and the specific wear rates (= material wear factors) on various high-temperature matrices at room temperature and T = 150°C. (From Lu, Z., Institut für Verbundwerkstoffe, Universität Kaiserslautern, Germany, 1994. With permission.)

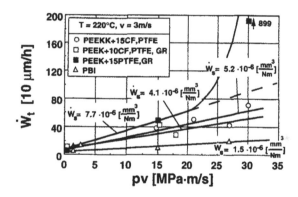

FIGURE 20
Specific wear rates up to high pressure times velocity conditions of various high-temperature polymers, especially modified for tribological applications at elevated temperatures, here, e.g., 220°C. (From Friedrich, K., Lu, Z., and Häger, A. M., *Theor. Appl. Fracture Mech.*, 19, 1, 1992. With permission.)

150 and 220°C (Figure 21). Self-lubrication was required in order not to make the textiles dirty by any grease or oil spots, and a low coefficient of friction was needed to operate the chain system under low-energy input. In summary, the following requirement profile should be fulfilled by the new high temperature polymer system to be found for this application:

1. Tribological properties $\mu \leq 0.1;\ \dot{w}_s \leq 10^{-6}\ \mathrm{mm^3/(Nm)}$
2. Temperature range $20°C \leq T \leq 220°C$
3. Mechanical performance $K_c \geq 5\ \mathrm{MPa}\ \sqrt{m}$ (fracture toughness)
4. Manufacturing aspects   Injection moldable into complex geometries

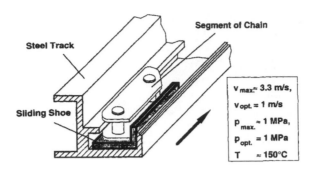

FIGURE 21
Schematic illustration of sliding shoes used in textile drying machines.

The high value for the fracture toughness was needed because the turn-around of the chains within the metallic guidance system could cause impact damage to the sliding shoes.

An evaluation of various potential material systems, as described in Figures 19 and 20, leads to the final choice of a PEEK matrix composite (including short carbon fibers, PTFE particles, and graphite flakes). In particular, this material was superior in the secondary material and the manufacturing requirements (Figure 22).[24] Under conditions of a major operating temperature of $T = 150°C$ and a pressure/velocity couple of p $= 0.1$ MPa/$v = 1$ m/sec the wear factor of this material of $k^* \triangleq \dot{w}_s = 1.11 \cdot 10^{-6}$ mm^3/(Nm) allowed to calculate via the relationship

$$\dot{w}_t = \dot{w}_s \cdot (p \cdot v)$$

with $\dot{w}_t$ depth wear rate [μm/h].

A maintenance interval of about 2.5 years assumed that the machine is operated under these conditions about 40 h/week (Table 2).

Result:

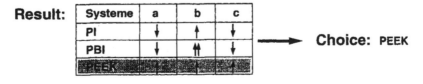

Systeme	a	b	c
PI	↓	↑	↓
PBI	↓	↑↑	↓
PEEK			

Choice: PEEK

FIGURE 22
Evaluation scheme of the matrix system most suitable for sliding shoes in textile drying machines. (From Lu, Z., Institut für Verbundwerkstoffe, Universität Kaiserslautern, Germany, 1994. With permission.)

TABLE 2.

Maintenance Intervals, Calculated from Specific Wear Rates of PEEK
Composites under Various External Loading Conditions

Temperature (°C)	Load (MPa)	Velocity (m/sec)	Δt Inspection (h)	In-Service Duration[a]
20	0.1	1	20,576	ca. 10 Years
($k_0^* = 0.27 \times 10^{-6}$	1	1	2,058	ca. 1 Year
[mm³/Nm])	0.1	3	6,859	ca. 3 Years
	1	3	1,686	ca. 4 Months
150	0.1	1	5,005	ca. 2.5 Years
($k_0^* = 1.11 \times 10^{-6}$	1	1	501	ca. 3 Months
[mm³/Nm])	0.1	3	1,668	ca. 10 Months
	1	3	167	ca. 1 Month

[a] In-service duration is calculated on the basis of 40 hours per week.

From Lu, Z., Institut für Verbundwerkstoffe, Universität Kaiserslautern, Germany, 1994. With permission.

C. Continuous Fiber/Polymer Matrix Composites

1. Systematic Development of Hybrid Composites

For a systematic development of a high-performance composite with optimum wear resistance, various unidirectional composites were tested with respect to their wear behavior as a function of matrix material, type of fiber reinforcement, and orientation of the fibers relative to the sliding direction. Various different matrices, e.g., a rather brittle epoxy resin and a very tough, high-temperature thermoplastics (PEEK), and three typical fiber materials, i.e., glass, carbon, and aramid, were chosen for this investigation (Table 3).[41-43] The results show quite systematically the same trends as outlined before:

1. The tougher PEEK matrix has a higher wear resistance than the brittle epoxy resin.
2. In the matrix with the lower wear rate, the beneficial effect of the fiber reinforcement on the composite's wear resistance is not as pronounced as in the easily wearing matrix.
3. For most of the fiber orientations carbon fibers give better results than glass fibers and aramid fibers are somewhere in between.

The wear resistances of these materials (inverse of the wear rates) can be utilized to systematically design a composite material with a generally good wear performance. An optimum wear resistance can be expected from a composite consisting of a PEEK matrix and carbon fibers parallel

TABLE 3.

Wear Resistance Ratios of Various UD Composites, as Normalized to that of the Neat Epoxy Resin

Matrix	Orientation	CF	GF	AF	Unreinforced
EP_1	N	65.4	0.5	122.4	1
	P	99.8	2.1	44.1	1
	AP	44.1	1.1	46.3	1
PEEK	N	65.1	5.1	61.4	11.8
	P	158.1	5.5	46.1	11.8
	AP	94.8	2.0	44.1	11.8
amPA	N	117.7	—	145.5	0.07
	P	128.6	—	65.1	0.07
	AP	57.0	—	102.4	0.07
PA66	N	41.3	1.8	145.5	27.7
	P	122.9	16.3	60.8	27.7
	AP	80.2	8.9	73.7	27.7
EP_2	N	102.4	—	104.3	1
	P	197.5	—	45.7	1
	AP	90.7	—	75.8	1

Note: EP: $\dot{w}_s^{-1} = 1.79 \cdot 10^4$ Nm/mm^3. The higher the ratio number, the better is the wear resistance, relative to the standard (EP).

to the sliding direction. If multidirectional sliding would be the case in the practical application, additional carbon fibers in the antiparallel orientation would probably be helpful. In fact, a study of such a material, PEEK with carbon fiber woven fabric reinforcement, resulted in even better wear factors than the two basic orientations of the unidirectional material. This synergistic effect is probably a result of fiber interlocking in the contact area due to the woven structure of the reinforcement. Further explanations are discussed in the article by Mody et al.[44]

A composite with an overall good wear resistance could be made by three-dimensional hybridization, with interwoven carbon fibers in plane (xy-plane) and aramid fibers in the z-direction. The matrix should be a very tough, high wear resistant polymer such as PEEK. Although such a three-dimensional hybrid structure could not be verified in the studies by Cirino et al.,[41] recent works with simpler two-dimensional hybrid composites point in the direction predicted. Krey et al.[45] have manufactured different laminates from amorphous polyamide (J2-polymer, DuPont, Wilmington, Delaware), impregnated carbon, and aramid fiber strands. Sliding these materials against rotating steel rings resulted in synergistic effects for the wear rates of hybrid composites with aramid fibers under normal orientation and carbon fibers under parallel orientation (Figure 23).

There are also some indications in the literature that composites with three-dimensional fiber arrangement and a high temperature polymer matrix with lubricants as inclusions have excellent wear resistance (Table 4).[46-48]

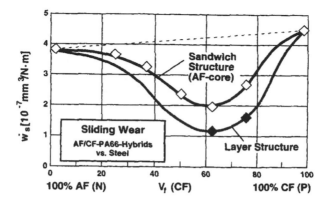

FIGURE 23

Synergistic effects for the wear resistance of two-dimensional hybrid composites with aramid fibers under normal orientation and carbon fibers under parallel orientation.

TABLE 4.

Depth Wear Rates of Three-Dimensional CF and GF-PI Composites Worn at Different Testing Temperatures

Fiber of Three-Dimensional Weave	Wear Rate (m/sec)		
	No Heat Applied	316°C	
	6.07 MPa · 0.76 m · s⁻¹	2.76 MPa · 0.51 s⁻¹	6.07 MPa · 0.76 m · s
CF-PI			
$\parallel$	$2.5 \cdot 10^{-8}$	$1.1 \cdot 10^{-8}$	$3.2 \cdot 10^{-8}$
$\perp$	$1.9 \cdot 10^{-8}$	$1.1 \cdot 10^{-8}$	$8.0 \cdot 10^{-8}$
GF-PI			
$\parallel$	$9.9 \cdot 10^{-9}$	$5.6 \cdot 10^{-9}$	$2.23 \cdot 10^{-7}$
$\perp$	$2.3 \cdot 10^{-8}$	$4.9 \cdot 10^{-9}$	$7.8 \cdot 10^{-8}$

$\parallel$ Majority of fiber lay is parallel with the plane of sliding.
$\perp$ Majority of fiber lay is normal to the plane of sliding.

From Gardos, M. N. and MacConnell, B. D., Development of a High-Load-Temperature Self-Lubricating Composite, Part I–IV, ASLE Prepr. 81-3A-3, 81-3A-6, 1981. With permission.

In the particular case mentioned, glass or carbon fibers were used as reinforcements. The three-dimensional structure was verified in different geometries of the sliding elements, e.g., as flat blocks and as circular bearings produced by a three-dimensional weaving technology with fibers in the radial, tangential, and axial direction.

2. Filament Wound Bearings of High Load Bearing Capacity

Apart from the more sophisticated, very complex and expensive three-dimensional woven structures, more simple, but rather successful filament winding technology has been used to produce polymer composite journal bearings on a commercial basis.[49,50] The cross sections of these bearings were characterized by various functions; i.e., in the inner range they were designed for low friction and low wear performance, whereas the outer structures were supposed to be resistant against creep and plastic deformation, so as to give the component a high load bearing capacity (Figure 24). Therefore, the advantages of these bearings were found to be in the range of high contact pressures, low velocities, and under dry lubrication or in extremely corrosive environments (Figure 25).[49]

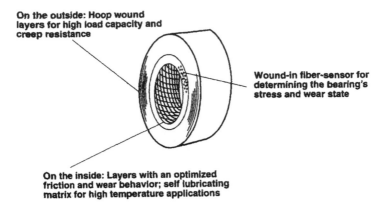

FIGURE 24
Schematic of a filament wound bearing.

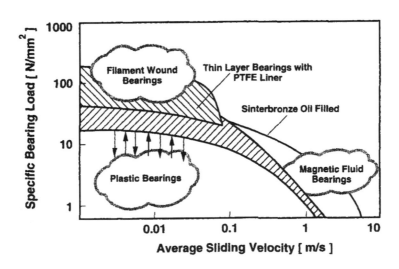

FIGURE 25
p-v-Diagram of various bearing materials, showing the high load bearing capacity of filament wound bearings at low sliding speed. (From K. Friedrich, as based on personal communication with SKF Gleitlager gmbH, Püttlingen, Germany, November 1992. With permission.)

A new group of filament wound bearings is under development at present, namely, those produced from continuous fiber reinforced, high-temperature composites. This requires, however, new technologies to be developed (in particular, with regard to the consolidation control) and the incorporation of sensor elements for preventing unexpected failure (i.e., wear control). Additional developments considered for the future are complete wear-resistant units, as produced by a combination of (1) filament winding of highly loadable bearing segments, and (2) the injection molding insert technique that allows for establishment of a more complex, wear-resistant structure around these segments (Figures 26).[51]

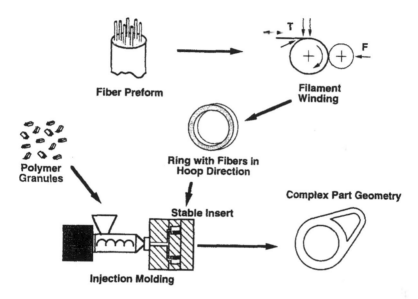

FIGURE 26
Schematic of the injection molding insert technique for manufacturing of complex, wear-resistant components. (From Haupert, F., Institut für Verbundwerkstoffe, Universität Kaiserlautern, July 5, 1994, 195. With permission.)

IV. CONCLUSIONS

This overview of the various possibilities for polymers and polymer composites to provide low friction and wear has necessarily been somewhat selective. Nevertheless, it is hoped that enough has been presented to show that there have been significant advances over the past decade or so, both in the development of improved composites and in the understanding of some aspects of their friction and wear mechanisms.

ACKNOWLEDGMENTS

These studies were supported by AGARD-Project G85 (collaboration with Professor S. A. Paipetis, Patras, Greece). Parts of the results have been previously published in *Composites Science and Technology*, 43, 71–84, 1992; in the Proceedings of Advanced Composites 1993, Wollongon, Australia, 1993; and in the Proceedings of AUSTRIB '94, Perth, Australia, December 1994. Further thanks are due to the Deutsche Forschungsgemeinschaft for research support on composite tribology (DFG-FR 675/19-1).

REFERENCES

1. Rohatgi, P. K., Blau, P. J., and Yust, C. S., Eds., *Tribology of Composite Materials, ASM International*, Materials Park, OH, 1990.
2. Zum Gahr, K.-H., *Microstructure and Wear of Materials*, Elsevier, Amsterdam, 1987.
3. Yamaguchi, Y., *Tribology of Plastic Materials*, Elsevier, Amsterdam, 1990.
4. Rymuza, Z., *Tribology of Miniature Systems*, Elsevier, Amsterdam, 1989.
5. Glaeser, W. A., *Materials for Tribology*, Elsevier, Amsterdam, 1992.
6. Bushan, B. and Gupta, B. K., *Handbook of Tribology, Materials, Coatings, and Surface Treatments*, McGraw-Hill, New York, 1991.
7. Dowson, D., Taylor, C. M., Godet, M., Eds., Mechanics of coatings, *Tribology Series*, Vol. 17, Elsevier, Amsterdam, 1990.
8. Stachowiak, G. W. and Batchelor, A. W., Engineering tribology, *Tribology Series*, Vol. 24, Elsevier, Amsterdam, 1993.
9. Lee, S. M., Ed., *International Encyclopedia of Composites*, Vol. 1–6, VCH Publishers, New York, 1990.
10. Friedrich, K., Ed., Friction and wear of polymer composites, *Composite Material Series*, Vol. 1, Elsevier, Amsterdam, 1986.
11. Friedrich, K. and Walter, R., Microstructure and tribological properties of short fiber thermoplastic composites, Proc. Tribology of Composite Materials, Oak Ridge, TN, May 1–3, 1990, 217.
12. Jacobs, O., Schulte, K., and Friedrich, K., Fretting wear and fretting fatigue of hybrid composites, in Proc. Int. Conf. on Composite Materials, ICCM VIII, Honolulu HI, July 1991, Vol. 4, 38 - G1 to 7.
13. Landheer, D. and Meesters, C. J. M., Reibung and Verschleiss von PTFE-Compounds, *Kunststoffe*, 83, 12, 996, 1993.
14. McLean, D. S., Tribological properties of acetal, nylon, and thermoplastic polyester, *Proc. ANTEC '94*, 1994, 3118.
15. Byett, J. H. and Allen, C., Dry Sliding Wear Behaviour of Polyamide 66 and Polycarbonate Composites, *Tribology Int.*, 25, no. 4, 237, 1992.
16. Böhm, H., Betz, S., and Ball, A., The wear resistance of polymers, *Tribol. Int.*, 23, no. 6, 399, 1990.
17. Beringer, H.-P., Heinke, G., and Strickle, E., Kunststoffe in der Tribotechnik — Polymere im Verschleisstest, *Tech. Rundsch.*, 25, 46, 1991.
18. McCamley, P. J., Selection guide to fiber-reinforced thermoplastics for high temperature application, *Adv. Mater. Proc.*, 8, 22, 1992.
19. LNP Eng. Plastics, Composite replaces aluminum in starter motor, *Adv. Mater. Proc.*, 7, 9, 1994.

20. Sandor, R. B., Polybenzimidazole (PBI) as a matrix resin precursor for carbon/carbon composites, *SAMPE Q.*, 4, 23, 1991.
21. Long, V. E., PBI performance parts: A business unit of Hoechst Celanese Corp. *Adv. Mater. Proc.*, 6, 50, 1992.
22. Münstedt, H. and Zeiner, H., Polyaryletherketone - Neue Möglichkeiten für Thermoplaste *Kunststoffe*, 79, 993, 1989.
23. Jacobs, O., Heitmann, T., Hoffmann, Th., and Wu, G. M., Sliding wear performance of drawn polyetheretherketone (PEEK), *Plastics Rubber Proc. Appl.*, 14, 1990, 203.
24. Lu, Z., Geschmierte Hochtemperatur-Verbundwerkstoffe für Anwendungen als Gleitelemente, Doktorarbeit, Institut für Verbundwerkstoffe, Universität Kaiserslautern, Germany, 1994.
25. Lancaster, J. K., Composites for increased wear resistance; current achievement and future prospects, NASA Conf. Publication Nr. 2300: Tribology in the 80's, Cleveland, April 18–21, 1983, Vol. 1, 1984, 333.
26. Häger, A. M. and Davies, M., Short fiber reinforced, high temperature resistant polymers for a wide field of tribological applications, in Friedrich, K., Ed., *Advances in Composite Tribology*, Elsevier Science Publ., Amsterdam, 1993, 107.
27. Friedrich, K., Ed., *Advances in Composite Tribology*, Elsevier Science Publ., Amsterdam, 1993.
28. Voss, H., Aufbau, Bruchverhalten und Verschleisseigenschaften kurzfaserverstärkter Hochleistungsthermoplaste, *Fortschrittsberichte VDI*, Reihe 5: Grund- und Werkstoffe, Nr. 116, VDI Verlag, Düsseldorf, Germany, 1987.
29. Friedrich, K., Karger-Kocsis, J., Sugioka, T., and Yoshida, M., On the sliding wear performance of polyethernitrile composites, *Wear*, 158, 157, 1992.
30. K. Friedrich, personal communication with Idemitsu Carbon Fiber, Idemitsu Kosan Co., Tokyo, Japan, 1991.
31. Luise, R. R., Liquid Crystalline Polymer Blends with Improved Wear Properties, International Patent Publication No.: WO 94/28069, May 25, 1994.
32. Schledjewski, R. and Friedrich, K., Tensile properties of filled liquid-crystal polymer systems, *J. Mater. Sci. Lett.*, 11, 840, 1992.
33. Voss, H. and Friedrich, K., Influence of short-fibre reinforcement on the fracture behavior of a bulk liquid crystal polymer, *J. Mater. Sci.*, 21, 2889, 1986.
34. Weng, T., Hiltner, A., and Baer, E., Hierarchical structure in a thermotropic liquid-crystalline copolyester, *J. Mater. Sci.*, 21, 744, 1986.
35. Kawabe, M., Yamaoka, I., and Kimura, M., Structures and thermal properties of liquid - crystalline poly(ester-co-carbonate) in *Liquid Crystalline Polymers*, Weiss, R. A. and Ober, C. K., Eds., ACS Symp. Ser., 435, chap. 9, 115.
36. Wu, J. S., Friedrich, K., and Grosso, M., Impact behavior of short fibre/liquid crystal polymer composites, *Composites*, 20, no. 3, 223, 1989.
37. Sawyer, L. C., Chen, R. T., Jamieson, M. G., Musselman, I. H., and Russell, P. E., the fibrillar hierarchy in liquid crystalline polymers, *J. Mater. Sci.*, 28, 225, 1993.
38. Kirsch, G. and Terwyen, H., Thermotrope Flüssigkristalline Polymere (LCP), *Kunststoffe*, 80, no. 10, 1159, 1990.
39. Schledjewski, R. and Friedrich, K., Fretting Wear Behavior of Filled LCP Systems Under Dry and Lubricated Conditions, in Proceedings of 13th International European Chapter Conference of the SAMPE, Hamburg, Germany, May 11–13, 1992, 93.
40. Friedrich, K., Lu, Z., and Häger, A. M., Overview on polymer composites for friction and wear application, *Theor. Appl. Fracture Mech.*, 19, 1, 1993.
41. Cirino, M., Friedrich, K., and Pipes, R. B., Evaluation of polymer composites for sliding and abrasive wear applications, *Composites*, 19, 5, 383, 1988.
42. Cirino, M., Pipes, R. B., and Friedrich, K., The abrasive wear behavior of continuous fiber polymer composites, *J. Mater. Sci.*, 22, 2481, 1987.
43. Cirino, M., Friedrich, K., and Pipes, R. B., The effect of fiber orientation on the abrasive wear behavior of polymer composite materials, *Wear*, 121, 127, 1988.

44. Mody, P. B., Chou, T. W., Friedrich, K., Effect of testing conditions and microstructure on the sliding wear of graphite fiber/PEEK matrix composites, *J. Mater. Sci.*, 23, 4319, 1988.

45. Krey, J., Friedrich, K., and Jacobs, O., Herstellung und Untersuchung der Verschleisseigenschaften von Hochleistungsverbundwerkstoffen mit Thermoplast matrix, *VDI-Berichte*, Nr. 737, 179, 1989.

46. Gardos, M. N. and MacConnell, B. D., Development of a High-Load-Temperature Self-Lubricating Composite, Part I–IV, ASLE-Prepr. 81-3A-3, 81-3A-6, 1981.

47. Gardos, M. N., Self-lubricating composites for extreme environmental conditions in Friedrich, K., Ed., *Friction and Wear of Polymer Composites*, Elsevier Science Publ., Amsterdam, 1986, 397.

48. Tribo-Comp TDF, Broschure, Tiodize Company, Huntington Beach, CA, 1985.

49. K. Friedrich, personal communication with SKF Gleitlager GmbH, Püttlingen, Germany, November 1992.

50. Garlock-Broschure: Gar-Fil and Gar-Max, Filament Wound, High Load Self-Lubricating Bearings Inc., Catalog 881, Garlock Bearings, Inc., Coltec Industries, U.S. 1992.

51. Haupert, F., Grundlagen der Verarbeitung thermoplastischer Verbundwerkstoffe, Festschrift Verbundwerkstoff-Kolloquium, Institut für Verbundwerkstoffe, Universität Kaiserslautern, July 5, 1994, 195.

I

A

Abrasion resistance, PC-HT, 217
Alkali salts, as catalysts, 30
Alloys, PBI, 175–176
Amic acid, 100
Amide thermosets, 82
Amorphous materials, moldings, 71–72
Apparent temperature index, for heat stress, 172–173
Appliances, 213–214
Applications
  aramid, 153–157
  PBI, 174–175
  polycarbonate (PC-HT), 212–218
  polyimide, 134–141
  TLCP, 16, 17, 22, 33, 57
Aramid, properties, 154
Aseptic packaging, 197
Aurum polyimide, 108, 159

B

Barrier properties, TLCP films, 19, 33, 46
Batteries, application, 175
Benzocyclobutene (BCB), dielectric, 140
Biaxial film, TLCP
  processing, 48, 49
  properties, 18, 44, 50, 51
Biochemical food degradation, 184
Bisacetylenes, LCT, 87
Bisphenol TMC, 205
Blends
  PBI fabric, 168–173
  PBI polymer, 175–176
  polyimide/TLCP, 116, 119
Burn behavior, PBI, 162
  protection, 163,
  severity, 165
  tests, mannequin, 171

C

Canned foods, 184
Carbon fiber, effect on wear, 231

Catalysis, propargyl thermosets, 91
Catalyst support, PBI, 176
Char formation
  aramids, 155–156
  PBI, 170
Chip I/O density, 133
Chronology, Patents, TLCP, 28
Circuit boards, printed, 124–127
Clothing, flame resistant, 155–158
Coalescence temperature, 103
Color
  aramids, 154
  polycarbonates, 210
  polyimides, 109
Composites, wear applications, 239–243
Computer electronics, 126–141
Conformational mobility, PC-HT, 204–205
Containers, TLCP, 54
Copolymers, Polyimide/TLCP, 113
Crosslinking, thermosets, 85, 128
Crystallization
  food packaging, 187–188
  heat-strengthening, 30
Curing kinetics, LCT, 87–88
Cut resistance, p-aramid fabrics, 157

D

Defects, injection-molding, 67–68
Die
  counter-rotating, 47
  trimodal, 52
Dielectrics, trends, 134–135
Dual ovenability, packaging, 198–199

E

Electrically conductive polymers, 143
Electrochemical applications, 174–175
Electronics
  multi-chip modules, 137
  packaging, computers, 131
  packaging, TLCP, 55

Elevated temperature wear,
    236–237
End groups, LCT, 84
Engineering plastics, molded
    properties, 73
Enzymatic food degradation, 184
Ester thermosets, 83

F

Fabrics, fire resistant, 166–173
Fibers, high temperature
    applications, 154–157
    aramid, 153–154
    melt-processible, 157
    polybenzimidazole (PBI), 166
    polyimide, 116–118, 120–121
    TLCP, 20, 29–31
Fibrids, aramid papers, 156
Filament wound bearings, wear,
    241–243
Filler, effect on wear
    carbon fiber, 231
    glass fiber, 232, 234
    PTFE, 227, 236
    TLCP wear, 235–236
Film, TLCP
    applications, 17, 53
    barrier properties, 19, 46
    co-extruded, 55
    extruded, 52
    heat-strengthened, 32
Filtration, hot-gas, 156
Fire protection, PBI, 163–165, 167
Flame protection, aramids, 154
Foams, TLCP, 37
Food, packaging, 181
    oven heating, 197–199
    product requirements, 183–184
Fretting wear, 228
Friction applications, 236–241
Fuel cells, 174

G

GE process, multi-chip modules, 138
Glass fiber alignment, moldings, 69–70
Glass transition temperature
    electronics, 124
    fibers, 151

PC-HT, 206
PET packaging, 188, 196
Graphite, effect on wear, 225

H

High-acid foods, packaging
    crystallization, 187–188
    hot-fill, 185–190
    sterilization, 190
Heat-ageing, polycarbonate (PC-HT),
    210
Heat-strengthening, TLCP
    conditions, 29
    crystallization, effect, 30
    mechanism, 30
    orientation, effect, 36
    patents, 27, 28
Hot-filling, packaging, 186
Hybid composites, wear applications,
    239–241
Hydrogen bonding, effect on melting
    point, 151
Hydrolyic stability, polyimides, 104
Hydroxybenzoic acid (HBA), 13

I

Ignition temperature, fabrics,
    164
Injection-molded, TLCP
    applications, 16
    heat-strengthened, 35
    properties, 16
    Scorim process, 69
Integrated circuits, 141
Interconnects, 141
Intermolecular forces, effect on
    melting point, 151

K

Kapton film, 103, 107
Kevlar fiber, 20, 32, 154

L

Ladder polymers, 153
LARC-IA polyimide, 112
LCT, definition, 80

Lens materials, polycarbonate, 215
Limiting oxygen index (LOI), 150
Logic chips, 129
Long-term heat applications, PC-HT, 213
Loss modulus, LCT, 90–91
Low-acid foods, packaging, 191–193
Lubricants, effect on polymer wear, 225, 228
Lyotropic polymers, 43, 153

M

Magnetic storage, electronics, 124
Magneto-optical applications, 218
Markets, TLCP, 22
Matrix, effect on polymer wear, 224–225, 227
Matrix resins, for printed circuit boards, 128
Medical applications
  TLCP, 20, 57
  polycarbonates, 216
Melt viscosity
  PC-HT, 207–208
  polyimides, 111
Metal can substitutes, 193–194
Microbiological food degradation, 184
Microwave susceptors, 200–202
Mid-term heat applications, PC-HT, 214–215
Moisture regain, PBI fabrics, 173
Molecular weight, polymer, effect on wear, 227
Molybdenum sulfide, effect on wear, 225
Multi-chip modules, 132, 138
Multi-layer boards, 56

N

Nanocomposites, LCT, 91–93
Naphthalenedicarboxylic acid (NDC), 7
Networks, liquid crystalline, 81
New-TPI, polyimide, 108–112
  see also Aurum, 159
Nomex fiber, 150

O

One-step synthesis, polyimides, 100
Optical path difference, PC-HT, 218
Optical storage, electronics, 125
Optoelectronics, 142–143
Oven heating, packaged foods, 197

P

Packaging, electronics, 124
Packaging, food, 181
Paper, fiber applications, 156
Patents, TLCP, 27–29
Perfluorocyclobutanes (PFCBs), dielectrics, 141
Phase diagram, LCT, 86, 94
Photolithography, 130
Photosensitive, polyimides, 137
Plastics, for packaging, 182
Polybenzimidazole (PBI), 152, 166
  fabrics, 169
  fiber properties, 167–168
  plastic parts, 175
Polycarbonate (PC-HT)
  high-heat applications
    long-term, 213
    mid-term, 214–215
    short-term, 217
  properties, 208–209, 211
Polycarbonates, for optical storage, 125
Polycyanurates, crosslinked network, 128
Poly(ester/imides), liquid crystalline, 116–118
Polyesters, liquid crystalline (TLCP)
  emerging developments, 4
  growth forecast, 5
  monomer supply, 13
  patent history, 27
  price analysis, 12
  suppliers, 9
  wear performance, 232–235
Polyetherketones, melt-spinnable, 158

Polyethylene, for packaging,
    182–183
Poly(ethylene naphthalate) (PEN)
    barrier film, 17–19
    packaging uses, 189
    wear performance, 231
Poly(ethylene terephthalate) (PET)
    barrier film, 17–19
    bottles, 187
    crystallization, 188, 196
    packaging uses, 183
Poly(imide/etherketones),
    114–115
Polyimides
    color, 109
    dielectrics, 139–140
    electronics packaging, 135–137
    liquid crystalline, 113, 116–119
    melt-processable, 108–112
    structural properties, 105
Polyisoimide, 101
Polyisoindoloquinoline (PIQ),
    dielectric, 139
Polymer wear
    filler, effect, 226, 227, 230–232
    lubricants, effect, 225, 228
    matrix, effect, 224–225, 227
Polyperfluoroethers, for electronics,
    125
Polypropylene,
    food packaging, 183
    moldings, 70
Poly(pyromellitimides)
    fibers, 152
    films, 105
Poly(vinyl alcohol), for packaging,
    183
Poly(vinylidene fluoride), for
    packaging, 183
Power dissipation, in chips, 131
Precision bearings, wear, 228
Printed circuit boards, 53–56,
    126–128
Processing head, Scorim, 63
Propargyl thermosets, 90
    catalysis, 91
Properties
    high temperature fibers, 150
    LCTs, 82–94
    polybenzimidazoles, 166

polycarbonates, 209–212
polyimides, 102–109, 116–118,
    120–121
TLCPs, 16, 18–20, 33–36, 44, 46,
    50–51, 150, 232–234
Protective clothing, 155–158
Pulp, TLCP, 37
Pultruded shapes, TLCP, 35

R

Reflector housings, 216
Reinforcements, effect on wear,
    225–231
Requirements, food packaging,
    183–184
Retort pouch, 193
Rheology, LCT, 90

S

Scorim process
    advantages, 66
    applications, 74–76
    properties, effect, 70–73
Semiconductors, 129–131
Short-term heat applications, PC-HT,
    217
Sliding shoes, wear application,
    236–238
Sliding wear, 223
Solid lubricants, effect on wear,
    225
Solubility, PC-HT, 212
Sterilization, food packaging, 190
Storage modulus
    LCT, 90–91
    PC-HT, 211
Structure
    high temperature fibers,
        151–153
    LCTs, 85–86
    TLCPs, 29–33, 44, 50–51
    polycarbonates, 207
    polyimides, 105
Surgical instruments, TLCP, 53, 57
Synthesis
    LCT, 81
    polycarbonates, 205
    polyimides, 100–102

T

Tenacity model, TLCP, 30
Ternary phase diagram, LCT, 94
Thermal stability, polyimides, 102
Thermosets, liquid crystalline
 (LCT)
  curing, 87–88
  phase behavior, 85–86
  thermal stability, 86–87
Thermosets, polyimide, 140–141
Thermotropic, 43
Three-dimensional composites, wear,
 240–241
TLCPs, definition, 4, 28
Transfer film formation, wear, 228
Transparent sheet, TLCP, 34
Trends, integrated circuits, 129
Tribology, 222
Trimodal die, 52
Tubing, TLCP, 33, 57
Two-step synthesis, polyimides,
 100–101

U

Uniaxial film, 45
Uniaxial sheet, 33
Upilex film, 107

V

Vapor deposition, polyimides, 102
Vectra TLCP resin, 7
Vectran TLCP fiber, 20, 32
Vertical burn test, PBI fabrics,
 169
Viscosity, melt-processible polyimide,
 111

W

Wear performance, TLCPs
 fillers, effect, 234–235
 mechanism, 233
 temperature, effect, 234
Wear rates, various polymers, 226
Wear testing, 223
Weathering, PC-HT, 210
Weld lines, injection-molding,
 75–76

X

Xydar TLCP resin, 7

Z

Zenite TLCP resin, 11

Printed and bound by CPI Group (UK) Ltd, Croydon, CR0 4YY

22/10/2024

01777622-0012